Multivariate Statistical Methods

A PRIMER

D1167174

Multivariate Statistical Methods

A PRIMER

SECOND EDITION

Bryan F.J. Manly

Department of Mathematics and Statistics
University of Otago
New Zealand

CHAPMAN & HALL

London · Glasgow · Weinheim · New York · Tokyo · Melbourne · Madras

Published by Chapman & Hall, 2-6 Boundary Row, London SE1 8HN, UK

Chapman & Hall, 2-6 Boundary Row, London SE1 8HN, UK

Blackie Academic & Professional, Wester Cleddens Road, Bishopbriggs, Glasgow G64 2NZ, UK

Chapman & Hall GmbH, Pappelallee 3, 69469 Weinheim, Germany

Chapman & Hall USA., 115 Fifth Avenue, New York, NY 10003, USA

Chapman & Hall Japan, ITP-Japan, Kyowa Building, 3F, 2-2-1 Hirakawacho, Chiyoda-ku, Tokyo 102, Japan

Chapman & Hall Australia, 102 Dodds Street, South Melbourne, Victoria 3205, Australia

Chapman & Hall India, R. Seshadri, 32 Second Main Road, CIT East, Madras 600 035, India

First edition 1986
Reprinted 1988, 1989, 1990, 1991, 1994

Second edition 1994
Reprinted 1995

© 1986, 1994 Bryan F.J. Manly

Typeset in 10/12 Times by Thomson Press (India) Ltd, New Delhi, India
Printed in Great Britain by St Edmundsbury Press, Bury St Edmunds, Suffolk

ISBN 0 412 60300 4

A Catalogue record for this book is available from the British Library

∞ Printed on permanent acid-free paper, manufactured in accordance with ANSI/NISO Z39.48-1992 and ANSI/NISO Z39.48-1984 (Permanence of Paper).

A journey of a thousand miles begins with a single step

Lao Tsu

Contents

Preface   **xi**

1   The material of multivariate analysis   **1**
  1.1 Examples of multivariate data   1
  1.2 Preview of multivariate methods   12
  1.3 The multivariate normal distribution   15
  1.4 Computer programs   16
  1.5 Graphical methods   16
      References   17

2   Matrix algebra   **18**
  2.1 The need for matrix algebra   18
  2.2 Matrices and vectors   18
  2.3 Operations on matrices   20
  2.4 Matrix inversion   22
  2.5 Quadratic forms   23
  2.6 Eigenvalues and eigenvectors   23
  2.7 Vectors of means and covariance matrices   24
  2.8 Further reading   26
      References   26

3   Displaying multivariate data   **27**
  3.1 The problem of displaying many variables
      in two dimensions   27
  3.2 Plotting index variables   27
  3.3 The draftsman's display   29
  3.4 The symbolic representation of data points   30
  3.5 Profiles of variables   32

3.6   Discussion and further reading   35
       References   36

4 Tests of significance with multivariate data   37
4.1   Simultaneous tests on several variables   37
4.2   Comparison of mean values for two samples:
       single variable case   37
4.3   Comparison of mean values for two samples:
       multivariate case   39
4.4   Multivariate versus univariate tests   43
4.5   Comparison of variation for two samples:
       single variable case   44
4.6   Comparison of variation for two samples:
       multivariate case   44
4.7   Comparisons of means for several samples   49
4.8   Comparison of variation for several samples   51
4.9   Computational methods and computer programs   53
       Exercise   53
       References   56

5 Measuring and testing multivariate distances   57
5.1   Multivariate distances   57
5.2   Distances between individual observations   57
5.3   Distances between populations and samples   62
5.4   Distances based upon proportions   67
5.5   Presence–absence data   68
5.6   The Mantel test   70
5.7   Computational methods and computer programs   73
5.8   Further reading   74
       Exercise   74
       References   75

6 Principal components analysis   76
6.1   Definition of principal components   76
6.2   Procedure for a principal components analysis   78
6.3   Computational methods and computer programs   88
6.4   Further reading   88
       Exercises   88
       References   91

Contents ix

7 Factor analysis **93**

7.1 The factor analysis model 93
7.2 Procedure for a factor analysis 95
7.3 Principal components factor analysis 97
7.4 Using a factor analysis program to do principal components analysis 99
7.5 Options in computer programs 104
7.6 The value of factor analysis 105
7.7 Computational methods and computer programs 105
7.8 Further reading 106
Exercise 106
References 106

8 Discriminant function analysis **107**

8.1 The problem of separating groups 107
8.2 Discrimination using Mahalanobis distances 108
8.3 Canonical discriminant functions 108
8.4 Tests of significance 110
8.5 Assumptions 112
8.6 Allowing for prior probabilities of group membership 117
8.7 Stepwise discriminant function analysis 117
8.8 Jackknife classification of individuals 118
8.9 Assigning of ungrouped individuals to groups 118
8.10 Logistic regression 118
8.11 Computational methods and computer programs 125
8.12 Further reading 126
Exercises 126
References 127

9 Cluster analysis **128**

9.1 Uses of cluster analysis 128
9.2 Types of cluster analysis 128
9.3 Hierarchic methods 130
9.4 Problems of cluster analysis 132
9.5 Measures of distance 133
9.6 Principal components analysis with cluster analysis 134
9.7 Computational methods and computer programs 140

9.8   Further reading   140
      Exercises   144
      References   144

10 Canonical correlation analysis   **146**

10.1 Generalizing a multiple regression analysis   146
10.2 Procedure for a canonical correlation analysis   148
10.3 Tests of significance   149
10.4 Interpreting canonical variates   151
10.5 Computational methods and computer programs   166
10.6 Further reading   166
      Exercise   167
      References   167

11 Multidimensional scaling   **168**

11.1 Constructing a 'map' from a distance matrix   168
11.2 Procedure for multidimensional scaling   170
11.3 Computational methods and computer programs   181
11.4 Further reading   181
      Exercise   182
      References   182

12 Ordination   **183**

12.1 The ordination problem   183
12.2 Principal components analysis   184
12.3 Principal coordinates analysis   190
12.4 Multidimensional scaling   198
12.5 Correspondence analysis   201
12.6 Comparison of ordination methods   205
12.7 Computational methods and computer programs   206
12.8 Further reading   206
      Exercise   206
      References   207

13 Epilogue   **208**

13.1 The next step   208
13.2 Some general reminders   208
13.3 Missing values   209
      References   210

Author index   **211**

Subject index   **213**

Preface

The purpose of this book is to introduce multivariate statistical methods to non-mathematicians. It is not intended to be comprehensive. Rather, the intention is to keep the details to a minimum while still conveying a good idea of what can be done. In other words, it is a book to 'get you going' in a particular area of statistical methods.

It is assumed that readers have a working knowledge of elementary statistics including tests of significance using the normal, t, chi-squared and F distributions, analysis of variance and linear regression. The material covered in a typical first-year university service course should be quite adequate, together with a reasonable facility in algebra. Describing multivariate statistical methods does require some use of matrix algebra. However, the amount needed is quite small if some details are accepted on faith and anyone who masters the material in Chapter 2 will have the required minimum level of competency in this area.

One of the reasons why multivariate methods are being used so often these days is the ready availability of computer programs to do the calculations. Indeed, access to suitable software is really essential if the methods are to be used. However, computer programs are not particularly stressed in this book because there are so many available that it is not realistic to write about them all or to concentrate on any particular one. What I have done is to mention the programs used to do the calculations for examples when this is appropriate, and in some cases to suggest other programs that also could have been used.

To a large extent the chapters can be read independently of each other. The first five are preliminary reading in that they are mainly concerned with general aspects of multivariate data rather than with specific techniques. Chapter 1 introduces some examples that are used in subsequent chapters and briefly describes the multivariate methods that the book is primarily concerned with. Chapter 2 covers matrix algebra and Chapter 3 is on graphical methods. Chapter 4 is about tests of significance. The material here is not crucial as far as understanding later chapters is concerned. Chapter 5 is about measuring distances using multivariate data. At least the first five sections should be read before Chapters 8 to 12.

Chapters 6 to 12 cover the most important multivariate methods in terms of current use. Of these, Chapter 12 on ordination is new in the second edition. Chapters 6 and 7 form a natural pair to be read together, and Chapter 12 draws on material in Chapters 6 and 11, but otherwise these chapters are fairly independent. Finally, in Chapter 13 I have attempted to sum up what has been covered and have made some general comments on good practice with the analysis of multivariate data.

The major changes from the first edition are (i) the introduction of the chapter on graphical methods in order to reflect the increased interest in this topic over the last few years, largely brought about by improvements in computer technology; (ii) more material on measuring distances between cases based on presence–absence data; (iii) a new section on the use of logistic regression as an alternative to discriminant function analysis with two groups; (iv) the introduction of a chapter on ordination, including sections on principal coordinates analysis and correspondence analysis; and (v) the introduction of exercises at the end of most chapters.

In making these changes I have continually kept in mind my original intention for the book, which was that it should be as short as possible and attempt to do no more than take readers to the stage where they can begin using multivariate methods in an intelligent manner.

I am indebted to many people for commenting on draft versions of the first edition. In particular, Earl Bardsley read early versions of several chapters, several anonymous reviewers read parts or all of the text and John Harraway read all of the final manuscript before it went to press. Their comments led to numerous improvements. Mary-Jane Campbell cheerfully typed and retyped the manuscript as I made changes, for which I am most grateful.

After the first edition appeared a number of readers have pointed out minor mistakes in the text so that I have been able to correct them in this edition. Liliana Gonzalez deserves a special mention for giving me a long list of suggested changes. Of course, I take full responsibility for any errors that still remain.

In conclusion I would like to thank the staff of Chapman and Hall for their work over the years in promoting the book, making small changes with reprints and encouraging me to produce this second edition.

Bryan F. J. Manly
Dunedin, November 1993

CHAPTER ONE

The material of multivariate analysis

1.1 Examples of multivariate data

The statistical methods that are described in elementary texts are mostly univariate methods because they are only concerned with analysing variation in a single random variable. This is even true of multiple regression because this technique involves trying to account for variation in one dependent variable. On the other hand, the whole point of a multivariate analysis is to consider several related random variables simultaneously, each one being considered equally important at the start of the analysis. The potential value of this more general approach is perhaps best seen by considering a few examples.

Example 1.1 Storm survival of sparrows

After a severe storm on 1 February 1898, a number of moribund sparrows were taken to the biological laboratory at Brown University, Rhode Island. Subsequently about half of the birds died and Hermon Bumpus saw this as an opportunity to study the effect of natural selection on the birds. He took eight morphological measurements on each bird and also weighed them. The results for five of the variables are shown in Table 1.1, for females only.

When Bumpus collected his data in 1898 his main interest was in the light that it would throw on Darwin's theory of natural selection. He concluded from studying the data that 'the birds which perished, perished not through accident, but because they were physically disqualified, and that the birds which survived, survived because they possessed certain physical characters'. To be specific, the survivors 'are shorter and weigh less...have longer wing bones, longer legs, longer sternums and greater brain capacity' than the non-survivors.

1

Table 1.1 Body measurements of female sparrows (X_1 = total length, X_2 = alar extent, X_3 = length of beak and head, X_4 = length of humerus, X_5 = length of keel of sternum; all in mm). Birds 1 to 21 survived, while the remainder died

Bird	X_1	X_2	X_3	X_4	X_5
1	156	245	31.6	18.5	20.5
2	154	240	30.4	17.9	19.6
3	153	240	31.0	18.4	20.6
4	153	236	30.9	17.7	20.2
5	155	243	31.5	18.6	20.3
6	163	247	32.0	19.0	20.9
7	157	238	30.9	18.4	20.2
8	155	239	32.8	18.6	21.2
9	164	248	32.7	19.1	21.1
10	158	238	31.0	18.8	22.0
11	158	240	31.3	18.6	22.0
12	160	244	31.1	18.6	20.5
13	161	246	32.3	19.3	21.8
14	157	245	32.0	19.1	20.0
15	157	235	31.5	18.1	19.8
16	156	237	30.9	18.0	20.3
17	158	244	31.4	18.5	21.6
18	153	238	30.5	18.2	20.9
19	155	236	30.3	18.5	20.1
20	163	246	32.5	18.6	21.9
21	159	236	31.5	18.0	21.5
22	155	240	31.4	18.0	20.7
23	156	240	31.5	18.2	20.6
24	160	242	32.6	18.8	21.7
25	152	232	30.3	17.2	19.8
26	160	250	31.7	18.8	22.5
27	155	237	31.0	18.5	20.0
28	157	245	32.2	19.5	21.4
29	165	245	33.1	19.8	22.7
30	153	231	30.1	17.3	19.8
31	162	239	30.3	18.0	23.1
32	162	243	31.6	18.8	21.3
33	159	245	31.8	18.5	21.7
34	159	247	30.9	18.1	19.0
35	155	243	30.9	18.5	21.3
36	162	252	31.9	19.1	22.2
37	152	230	30.4	17.3	18.6
38	159	242	30.8	18.2	20.5
39	155	238	31.2	17.9	19.3
40	163	249	33.4	19.5	22.8

Table 1.1 (*Contd.*)

Bird	X_1	X_2	X_3	X_4	X_5
41	163	242	31.0	18.1	20.7
42	156	237	31.7	18.2	20.3
43	159	238	31.5	18.4	20.3
44	161	245	32.1	19.1	20.8
45	155	235	30.7	17.7	19.6
46	162	247	31.9	19.1	20.4
47	153	237	30.6	18.6	20.4
48	162	245	32.5	18.5	21.1
49	164	248	32.3	18.8	20.9

Data source: Bumpus (1898).

He also concluded that 'the process of selective elimination is most severe with extremely variable individuals, no matter in which direction the variations may occur. It is quite as dangerous to be conspicuously above a certain standard of organic excellence as it is to be conspicuously below the standard.' This last statement is saying that stabilizing selection occurred, so that individuals with measurements close to the average survived better than individuals with measurements rather different from the average.

Of course, the development of multivariate statistical methods had hardly begun in 1898 when Bumpus was writing. The correlation coefficient as a measure of the relationships between two variables was introduced by Francis Galton in 1877. However, it was another 56 years before Hotelling described a practical method for carrying out a principal components analysis, which is one of the simplest multivariate analyses that can be applied to Bumpus's data. In fact Bumpus did not even calculate standard deviations. Nevertheless, his methods of analysis were sensible. Many authors have reanalysed his data and, in general, have confirmed his conclusions.

Taking the data an an example for illustrating multivariate techniques, several interesting questions spring to mind. In particular:

1. How are the different measurements related? For example, does a large value for one variable tend to occur with large values for the other variables?
2. Do the survivors and non-survivors have significant differences for the mean values of the variables?

Table 1.2 Measurements on male Egyptian skulls from various epochs (X_1 = maximum breadth, X_2 = basibregmatic height, X_3 = basialveolar length, X_4 = nasal height; all in mm, as shown on Fig. 1.1)

Skull	Early predynastic				Late predynastic				12th & 13th dynasties				Ptolemaic period				Roman period			
	X_1	X_2	X_3	X_4	X_1	X_2	X_3	X_4	X_1	X_2	X_3	X_4	X_1	X_2	X_3	X_4	X_1	X_2	X_3	X_4
1	131	138	89	49	124	138	101	48	137	141	96	52	137	134	107	54	137	123	91	50
2	125	131	92	48	133	134	97	48	129	133	93	47	141	128	95	53	136	131	95	49
3	131	132	99	50	138	134	98	45	132	138	87	48	141	130	87	49	128	126	91	57
4	119	132	96	44	148	129	104	51	130	134	106	50	135	131	99	51	130	134	92	52
5	136	143	100	54	126	124	95	45	134	134	96	45	133	120	91	46	138	127	86	47
6	138	137	89	56	135	136	98	52	140	133	98	50	131	135	90	50	126	138	101	52
7	139	130	108	48	132	145	100	54	138	138	95	47	140	137	94	60	136	138	97	58
8	125	136	93	48	133	130	102	48	136	145	99	55	139	130	90	48	126	126	92	45
9	131	134	102	51	131	134	96	50	136	131	92	46	140	134	90	51	132	132	99	55
10	134	134	99	51	133	125	94	46	126	136	95	56	138	140	100	52	139	135	92	54
11	129	138	95	50	133	136	103	53	137	129	100	53	132	133	90	53	143	120	95	51
12	134	121	95	53	131	139	98	51	137	139	97	50	134	134	97	54	141	136	101	54
13	126	129	109	51	131	136	99	56	136	126	101	50	135	135	99	50	135	135	95	56
14	132	136	100	50	138	134	98	49	137	133	90	49	133	136	95	52	137	134	93	53
15	141	140	100	51	130	136	104	53	129	142	104	47	136	130	99	55	142	135	96	52

16	131	134	97	54	131	128	98	45	135	138	102	55	134	137	93	52	139	134	95	47
17	135	137	103	50	138	129	107	53	129	135	92	50	131	141	99	55	138	125	99	51
18	132	133	93	53	123	131	101	51	134	125	90	60	129	135	95	47	137	135	96	54
19	139	136	96	50	130	129	105	47	138	134	96	51	136	128	93	54	133	125	92	50
20	132	131	101	49	134	130	93	54	136	135	94	53	131	125	88	48	145	129	89	47
21	126	133	102	51	137	136	106	49	132	130	91	52	139	130	94	53	138	136	92	46
22	135	135	103	47	126	131	100	48	133	131	100	50	144	124	86	50	131	129	97	44
23	134	124	93	53	135	136	97	52	138	137	94	51	141	131	97	53	143	126	88	54
24	128	134	103	50	129	126	91	50	130	127	99	45	130	131	98	53	134	124	91	55
25	130	130	104	49	134	139	101	49	136	133	91	49	133	128	92	51	132	127	97	52
26	138	135	100	55	131	134	90	53	134	123	95	52	138	126	97	54	137	125	85	57
27	128	132	93	53	132	130	104	50	136	137	101	54	131	142	95	53	129	128	81	52
28	127	129	106	48	130	132	93	52	133	131	96	49	136	138	94	55	140	135	103	48
29	131	136	114	54	135	132	98	54	138	133	100	55	132	136	92	52	147	129	87	48
30	124	138	101	46	130	128	101	51	138	133	91	46	135	130	100	51	136	133	97	51

3. Do survivors and non-survivors show the same amount of variation in measurements?

4. If the survivors and non-survivors differ with regard to their distributions for the variables, is it possible to construct some function of these variables $f(X_1, X_2, X_3, X_4, X_5)$ which separates the two groups? It would be convenient if this function tended to be large for survivors and small for non-survivors since it would then be an index of Darwinian fitness.

Example 1.2 Egyptian skulls

For a second example, consider the data shown in Table 1.2 for measurements made on male Egyptian skulls from the area of Thebes. There are five samples of 30 skulls from each of the early predynastic period (*circa* 4000 BC), the late predynastic period (*circa* 3300 BC), the 12th and 13th dynasties (*circa* 1850 BC), the Ptolemaic period (*circa* 200 BC), and the Roman period (*circa* AD 150). Four measurements are available on each skull, as shown in Fig. 1.1.

1. How are the four measurements related?

2. Are there significant differences in the sample means for the variables and, if so, do these differences reflect gradual changes with time?

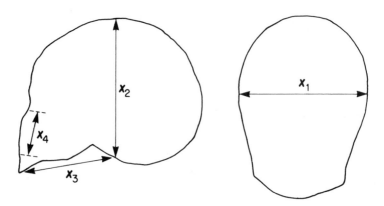

Figure 1.1 Measurements on Egyptian skulls.

3. Are there significant differences in the sample standard deviations for the variables and, if so, do the differences reflect gradual changes with time?
4. Is it possible to construct a function $f(X_1, X_2, X_3, X_4)$ of the four variables that in some sense captures most of the sample differences?

These questions are, of course, rather similar to the one suggested with Example 1.1.

As will be seen later, there are differences between the five samples that can be explained partly as time trends. It must be said, however, that the reasons for the changes are not known. Migration into the population was probably the most important factor.

Example 1.3 Distribution of a butterfly

A study of 16 colonies of the butterfly *Euphydryas editha* in California and Oregon produced the data shown in Table 1.3. Here there are two types of variable: environmental and biological. The environmental variables are altitude, rainfall, and minimum and maximum temperatures. The biological variables are gene frequencies for phosphoglucose-isomerase (Pgi) as determined by the technique of electrophoresis. For the present purposes there is no need to go into the details of how the gene frequencies were determined. (Strictly speaking these are not gene frequencies anyway.) It is sufficient to say that the frequencies describe the genetic distribution of *E. editha* to some extent. Figure 1.2 shows the geographical positions of the colonies.

In this example questions that can be asked are:

1. Are the Pgi frequencies similar for colonies that are close in space?
2. To what extent are the Pgi frequencies related to the environmental variables?

These questions are important when it comes to trying to decide how Pgi frequencies are determined. If frequencies are largely determined by present and past migration then gene frequencies should be similar for adjacent colonies but may show no relationships with environmental variables. On the other hand, if it is the environment that is most important then the Pgi frequencies should be related to the environmental variables, but colonies that are close in space

Table 1.3 Environmental variables and phosphoglucose-isomerase (Pgi) gene frequencies for colonies of the butterfly *Euphydryas editha* in California and Oregon

Colony	Altitude (feet)	Annual precipitation (inches)	Annual maximum temperature (°F)	Annual minimum temperature (°F)	Pgi mobility gene frequencies (%)					
					0.40	0.60	0.80	1.00	1.16	1.30
SS	500	43	98	17	0	3	22	57	17	1
SB	800	20	92	32	0	16	20	38	13	13
WSB	570	28	98	26	0	6	28	46	17	3
JRC	550	28	98	26	0	4	19	47	27	3
JRH	550	28	98	26	0	1	8	50	35	6
SJ	380	15	99	28	0	2	19	44	32	3
CR	930	21	99	28	0	0	15	50	27	8
UO	650	10	101	27	10	21	40	25	4	0
LO	600	10	101	27	14	26	32	28	0	0
DP	1 500	19	99	23	0	1	6	80	12	1
PZ	1 750	22	101	27	1	4	34	33	22	6
MC	2 000	58	100	18	0	7	14	66	13	0
IF	2 500	34	102	16	0	9	15	47	21	8
AF	2 000	21	105	20	3	7	17	32	27	14
GH	7 850	42	84	5	0	5	7	84	4	0
GL	10 500	50	81	−12	0	3	1	92	4	0

Data source: McKechnie *et al.* (1975). Environmental variables have been rounded to integers for simplicity. The original data were for 21 colonies. For the present example, five colonies with small samples for the estimation of gene frequencies have been excluded so as to make all of the estimates about equally reliable.

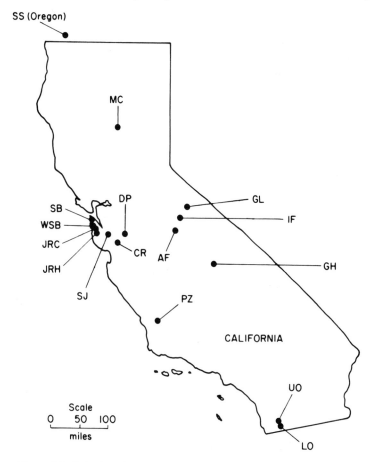

Figure 1.2 Colonies of *Euphydryas editha* in California and Oregon.

will have different frequencies if the environments are different. Of course, colonies that are close in space will tend to have similar environments. It may therefore be difficult to reach a clear conclusion.

Example 1.4 Prehistoric dogs from Thailand

Excavations of prehistoric sites in northeast Thailand have produced a series of canid (dog) bones covering a period from about 3500 BC to the present. The origin of the prehistoric dog is not certain. It

Table 1.4 Mean mandible measurements (mm) for modern Thai dogs, golden jackals, wolves, cuons, dingos and prehistoric dogs (X_1 = breadth of mandible, X_2 = height of mandible below 1st molar, X_3 = length of 1st molar, X_4 = breadth of 1st molar, X_5 = length from 1st to 3rd molars inclusive, X_6 = length from 1st to 4th premolars inclusive)

Group	X_1	X_2	X_3	X_4	X_5	X_6
Modern dog	9.7	21.0	19.4	7.7	32.0	36.5
Golden jackal	8.1	16.7	18.3	7.0	30.3	32.9
Chinese wolf	13.5	27.3	26.8	10.6	41.9	48.1
Indian wolf	11.5	24.3	24.5	9.3	40.0	44.6
Cuon	10.7	23.5	21.4	8.5	28.8	37.6
Dingo	9.6	22.6	21.1	8.3	34.4	43.1
Prehistoric dog	10.3	22.1	19.1	8.1	32.3	35.0

Data source: Higham *et al.* (1980).

could descend from the golden jackal (*Canis aureus*) or the wolf. However, the wolf is not native to Thailand, the nearest indigenous sources being western China (*Canis lupus chanco*) or the Indian subcontinent (*Canis lupus pallipes*).

In order to clarify the ancestry of the prehistoric dogs, mandible measurements were made on the available specimens. These were then compared with similar measurements on the golden jackal, the Chinese wolf and the Indian wolf. The comparisons were made more useful by considering also the dingo, which may have its origins in India, the cuon (*Cuon alpinus*) which is indigenous to southeast Asia, and modern village dogs from Thailand.

Table 1.4 gives mean values for six of the mandible measurements for specimens from all of the groups. The main question to be addressed here is how these groups are related and, in particular, how the prehistoric group is related to the others.

Example 1.5 Employment in European countries

Finally, as a contrast to the previous biological examples, consider the data in Table 1.5. This shows the percentages of the labour force in nine different types of industry for 26 European countries. In this case multivariate analyses may be useful in isolating groups of countries with similar employment distributions and in generally aiding the comprehension of the relationships between the countries.

Table 1.5 Percentages of people employed in nine different industry groups in Europe (AGR = agriculture, MIN = mining, MAN = manufacturing, PS = power supplies, CON = construction, SER = service industries, FIN = finance, SPS = social and personal services, TC = transport and communications)

Country	AGR	MIN	MAN	PS	CON	SER	FIN	SPS	TC
Belgium	3.3	0.9	27.6	0.9	8.2	19.1	6.2	26.6	7.2
Denmark	9.2	0.1	21.8	0.6	8.3	14.6	6.5	32.2	7.1
France	10.8	0.8	27.5	0.9	8.9	16.8	6.0	22.6	5.7
W. Germany	6.7	1.3	35.8	0.9	7.3	14.4	5.0	22.3	6.1
Ireland	23.2	1.0	20.7	1.3	7.5	16.8	2.8	20.8	6.1
Italy	15.9	0.6	27.6	0.5	10.0	18.1	1.6	20.1	5.7
Luxembourg	7.7	3.1	30.8	0.8	9.2	18.5	4.6	19.2	6.2
Netherlands	6.3	0.1	22.5	1.0	9.9	18.0	6.8	28.5	6.8
UK	2.7	1.4	30.2	1.4	6.9	16.9	5.7	28.3	6.4
Austria	12.7	1.1	30.2	1.4	9.0	16.8	4.9	16.8	7.0
Finland	13.0	0.4	25.9	1.3	7.4	14.7	5.5	24.3	7.6
Greece	41.4	0.6	17.6	0.6	8.1	11.5	2.4	11.0	6.7
Norway	9.0	0.5	22.4	0.8	8.6	16.9	4.7	27.6	9.4
Portugal	27.8	0.3	24.5	0.6	8.4	13.3	2.7	16.7	5.7
Spain	22.9	0.8	28.5	0.7	11.5	9.7	8.5	11.8	5.5
Sweden	6.1	0.4	25.9	0.8	7.2	14.4	6.0	32.4	6.8
Switzerland	7.7	0.2	37.8	0.8	9.5	17.5	5.3	15.4	5.7
Turkey	66.8	0.7	7.9	0.1	2.8	5.2	1.1	11.9	3.2
Bulgaria	23.6	1.9	32.3	0.6	7.9	8.0	0.7	18.2	6.7
Czechoslovakia	16.5	2.9	35.5	1.2	8.7	9.2	0.9	17.9	7.0
E. Germany	4.2	2.9	41.2	1.3	7.6	11.2	1.2	22.1	8.4
Hungary	21.7	3.1	29.6	1.9	8.2	9.4	0.9	17.2	8.0
Poland	31.1	2.5	25.7	0.9	8.4	7.5	0.9	16.1	6.9
Romania	34.7	2.1	30.1	0.6	8.7	5.9	1.3	11.7	5.0
USSR	23.7	1.4	25.8	0.6	9.2	6.1	0.5	23.6	9.3
Yugoslavia	48.7	1.5	16.8	1.1	4.9	6.4	11.3	5.3	4.0

Source: Euromonitor (1979, pp. 76–7) with the percentage employed in finance in Spain reduced from 14.7 to the more reasonable figure of 8.5.

1.2 Preview of multivariate methods

The five examples just considered are typical of the raw material for multivariate statistical methods. The main thing to note at this point is that in all cases there are several variables of interest and these are clearly not independent of each other. However, it is useful also to give a brief preview of what is to come in the chapters that follow in relationship to these examples.

Principal components analysis is designed to reduce the number of variables that need to be considered to a small number of indices (called the principal components) that are linear combinations of the original variables. For example, much of the variation in the body measurements of sparrows shown in Table 1.1 will be related to the general size of the birds, and the total

$$I_1 = X_1 + X_2 + X_3 + X_4 + X_5$$

will measure this quite well. This accounts for one 'dimension' in the data. Another index is

$$I_2 = X_1 + X_2 + X_3 - X_4 - X_5,$$

which is a contrast between the first three measurements and the last two. This reflects another 'dimension' in the data. Principal components analysis provides an objective way of finding indices of this type so that the variation in the data can be accounted for as concisely as possible. It may well turn out that two or three principal components provide a good summary of all the original variables. Consideration of the values of the principal components instead of the values of the original variables may then make it much easier to understand what the data have to say. In short, principal components analysis is a means of simplifying data by reducing the number of variables.

Factor analysis also attempts to account for the variation in a number of original variables using a smaller number of index variables or factors. It is assumed that each original variable can be expressed as a linear combination of these factors, plus a residual term that reflects the extent to which the variable is independent of the other variables. For example, a two-factor model for the sparrow data assumes that

$$X_1 = a_{11}F_1 + a_{12}F_2 + e_1$$
$$X_2 = a_{21}F_1 + a_{22}F_2 + e_2$$
$$X_3 = a_{31}F_1 + a_{32}F_2 + e_3$$
$$X_4 = a_{41}F_1 + a_{42}F_2 + e_4$$

and

$$X_5 = a_{51}F_1 + a_{52}F_2 + e_5,$$

where the a_{ij} values are constants, F_1 and F_2 are the factors, and e_i represents the variation in X_i that is independent of the variation in the other X-variables. Here F_1 might be the factor of size. In that case the coefficients a_{11}, a_{21}, a_{31}, a_{41} and a_{51} would all be positive, reflecting the fact that some birds tend to be large and some birds tend to be small on all body measurements. The second factor F_2 might then measure an aspect of the shape of birds, with some positive coefficients and some negative coefficients. If this two-factor model fitted the data well then it would provide a relatively straightforward description of the relationship between the five body measurements being considered.

One type of factor analysis starts by taking a few principal components as the factors in the data being considered. These initial factors are then modified by a special transformation process called 'factor rotation' in order to make them easier to interpret. Other methods for finding initial factors are also used. A rotation to simpler factors is almost always done.

Discriminant function analysis is concerned with the problem of seeing whether it is possible to separate different groups on the basis of the available measurements. This could be used, for example, to see how well surviving and non-surviving sparrows can be separated using their body measurements (Example 1.1), or how skulls from different epochs can be separated, again using size measurements (Example 1.2). Like principal components analysis, discriminant function analysis is based on the idea of finding suitable linear combinations of the original variables.

Cluster analysis is concerned with the identification of groups of similar objects. There is not much point in doing this type of analysis with data like those of Example 1.1 and 1.2 as the groups (survivors, non-survivors; epochs) are already known. However, in Example 1.3 there might be some interest in grouping colonies on the basis of environmental variables or Pgi frequencies, while in Example 1.4 the main point of interest is in the similarity between prehistoric dogs and other animals. Likewise, in Example 1.5 different European countries can be grouped in terms of similarity between employment patterns.

With *canonical correlation* the variables (not the objects) are divided into two groups and interest centres on the relationship between these. Thus in Example 1.3 the first four variables are related to the environment while the remaining six variables reflect the genetic distribution at the different colonies of *Euphydryas editha*. Finding what relationships, if any, exist between these two groups of variables is of considerable biological interest.

Multidimensional scaling begins with data on some measure of the distances apart of a number of objects. From these distances a 'map' is constructed showing how the objects are related. This is a useful facility as it is often possible to measure how far apart pairs of objects are without having any idea of how the objects are related in a geometric sense. Thus in Example 1.4 there are ways of measuring the 'distances' between modern dogs and golden jackals, modern dogs and Chinese wolves, etc. Considering each pair of animal groups gives 21 distances altogether. From these distances multidimensional scaling can be used to produce a 'map' of the relationships between the groups. With a one-dimensional 'map' the groups are placed along a straight line. With a two-dimensional 'map' they are represented by points on a plane. With a three-dimensional 'map' they are represented by points within a cube. Four- and higher-dimensional solutions are also possible although these have limited use because they cannot be visualized. The value of a one-, two- or three-dimensional map is clear for Example 1.4 as such a map would immediately show which groups prehistoric dogs are similar to and which groups they are different from. Hence multidimensional scaling may be a useful alternative to cluster analysis in this case. A 'map' of European countries by employment patterns might also be of interest in Example 1.5.

Principal components analysis and multidimensional scaling are sometimes referred to as methods for *ordination*. That is to say, they are methods for producing axes against which a set of objects of interest can be plotted. Other methods of ordination are also available, in particular *principal coordinates analysis* and *correspondence analysis.*

Principal coordinates analysis is like a type of principal components analysis that starts off with information on the extent to which a set of objects are different instead of the values for measurements on the objects. As such, it is intended to do the same as multi-dimensional scaling. However the assumptions made and the numerical methods used are not the same.

Correspondence analysis is a method for constructing axes against which to simultaneously plot both the objects of interest and the characteristics that are used to describe the objects starting with data on the abundance of each of the characteristics for each of the objects. This is useful in ecology, for example, where the objects of interest are often different sites, the characteristics are different species, and the data consist of abundances of the species in samples taken from the sites. The purpose of correspondence analysis would then be to clarify the relationships between the sites as expressed by species distributions, and the relationships between the species as expressed by site distributions. The method is popular in this context because the sites and species can be plotted together to give what may be a very informative ordination.

1.3 The multivariate normal distribution

The normal distribution for a single variable should be familiar to readers of this book. It has the well-known 'bell-shaped' frequency curve. Many standard univariate statistical methods are based on the assumption that data are normally distributed.

Knowing the prominence of the normal distribution with univariate statistical methods, it will come as no surprise to discover that the multivariate normal distribution has a central position with multivariate statistical methods. Many of these methods require the assumption that the data being analysed have multivariate normal distributions.

The exact definition of a multivariate normal distribution is not too important. The approach of most people, for better or worse, seems to be to regard data as being normally distributed unless there is some reason to believe that this is not true. In particular, if all the individual variables being studied appear to be normally distributed, then it is assumed that the joint distribution is multivariate normal. This is, in fact, a minimum requirement since the definition of multivariate normality requires more than this.

Cases do arise where the assumption of multivariate normality is clearly invalid. For example, one or more of the variables being studied may have a highly skewed distribution with several outlying high (or low) values; there may be many repeated values, etc. This type of problem can sometimes be overcome by an appropriate transformation of the data, as discussed in elementary texts on statistics. If this does not work then a rather special form of analysis may be required.

One important aspect of a multivariate normal distribution is that it is specified completely by a mean vector and a covariance matrix. The definitions of a mean vector and a covariance matrix are given in section 2.7.

1.4 Computer programs

Practical methods for carrying out the calculations for multivariate analyses have been developed over the last 60 years or so. However, the application of these methods for more than a small number of variables had to wait until computers became available. Therefore, it is only in the last 20 years that the methods have become reasonably easy to carry out for the average researcher.

Nowadays there are many standard statistical packages and computer programs available for calculations on large mainframe computers, work stations and personal computers. It is the intention that this book should provide readers with enough information to use any of these packages and programs intelligently, without saying much about any particular one. However, where it is appropriate the software used to analyse data will be mentioned in the examples in the following chapters.

1.5 Graphical methods

One of the outcomes of the greatly improved computer facilities in recent times has been an increase in the variety of graphical methods for multivariate data that are available. This includes contour plots and three-dimensional surface plots for functions of two variables, and a variety of special methods for showing the values that individual cases have for three or more variables. These methods are becoming used more commonly as part of the analysis of multivariate data and they are therefore discussed at some length in Chapter 3.

References

Bumpus, H.C. (1898) The elimination of the unfit as illustrated by the introduced sparrow, *Passer domesticus*. *Biological Lectures, Marine Biology Laboratory, Woods Hole*, 11th Lecture, pp. 209–26.

Euromonitor (1979) *European Marketing Data and Statistics*. Euromonitor Publications, London.

Higham, C.F.W., Kijngam, A. and Manly, B.F.J. (1980) An analysis of prehistoric canid remains from Thailand. *Journal of Archaeological Science* 7, 149–65.

McKechnie, S.W., Ehrlich, P.R. and White, R.R. (1975) Population genetics of *Euphydryas* butterflies. I. Genetic variation and the neurality hypothesis. *Genetics* 81, 571–94.

Thomson, A. and Randall-Maciver, R. (1905) *Ancient Races of the Thebaid*. Oxford University Press.

CHAPTER TWO

Matrix algebra

2.1 The need for matrix algebra

As indicated in the Preface, the theory of multivariate statistical methods can only be explained reasonably well with the use of some matrix algebra. For this reason it is helpful, if not essential, to have a certain minimal knowledge of this area of mathematics. This is true even for those whose interest is solely in using the methods as a tool. At first sight, the notation of matrix algebra is certainly somewhat daunting. However, it is not difficult to understand the basic principles involved providing that some details are accepted on faith.

2.2 Matrices and vectors

A *matrix* of size $m \times n$ is an array of numbers with m rows and n columns, considered as a single entity, of the form

$$
\mathbf{A} = \begin{bmatrix} a_{11} & a_{12} & \cdots & a_{1n} \\ a_{21} & a_{22} & \cdots & a_{2n} \\ \vdots & \vdots & & \\ a_{m1} & a_{m2} & \cdots & a_{mn} \end{bmatrix}.
$$

If $m = n$, then this is a *square* matrix. If a matrix only has one column, for instance,

$$
\mathbf{c} = \begin{bmatrix} c_1 \\ c_2 \\ \vdots \\ c_m \end{bmatrix},
$$

18

then this is called a *column vector*. If there is only one row, for instance

$$\mathbf{r} = (r_1, r_2, \ldots, r_n),$$

then this is called a *row vector*.

The *transpose* of a matrix is found by interchanging the rows and columns. Thus the transpose of **A** above is

$$\mathbf{A}' = \begin{bmatrix} a_{11} & a_{21} & \cdots & a_{m1} \\ a_{12} & a_{22} & \cdots & a_{m2} \\ \vdots & \vdots & & \\ a_{1n} & a_{2n} & \cdots & a_{mn} \end{bmatrix}.$$

Also, $\mathbf{c}' = (c_1, c_2, \ldots, c_m)$, and $\mathbf{r}'$ is a column vector.

There are a number of special kinds of matrix that are particularly important. A *zero matrix* has all elements equal to zero:

$$\mathbf{0} = \begin{bmatrix} 0 & 0 & \cdots & 0 \\ 0 & 0 & \cdots & 0 \\ \vdots & & & \\ 0 & 0 & \cdots & 0 \end{bmatrix}.$$

A *diagonal matrix* is a square matrix that has elements of zero, except down the main diagonal, and so is of the form

$$\mathbf{T} = \begin{bmatrix} t_1 & 0 & 0 & \cdots & 0 \\ 0 & t_2 & 0 & \cdots & 0 \\ 0 & 0 & t_3 & \cdots & 0 \\ \vdots & & & & \\ 0 & 0 & 0 & \cdots & t_n \end{bmatrix}.$$

A *symmetric matrix* is a square matrix that is unchanged when it is transposed, so that **A** has this property providing that $\mathbf{A}' = \mathbf{A}$. Finally, an *identity matrix* is a diagonal matrix with all diagonal terms being unity:

$$\mathbf{T} = \begin{bmatrix} 1 & 0 & 0 & \cdots & 0 \\ 0 & 1 & 0 & \cdots & 0 \\ 0 & 0 & 1 & \cdots & 0 \\ \vdots & \vdots & \vdots & & \vdots \\ 0 & 0 & 0 & \cdots & 1 \end{bmatrix}.$$

Two matrices are *equal* only if all their elements agree. For example,

$$\begin{bmatrix} a_{11} & a_{12} & a_{13} \\ a_{21} & a_{22} & a_{23} \\ a_{31} & a_{32} & a_{33} \end{bmatrix} = \begin{bmatrix} b_{11} & b_{12} & b_{13} \\ b_{21} & b_{22} & b_{23} \\ b_{31} & b_{32} & b_{33} \end{bmatrix}$$

only if $a_{11} = b_{11}$, $a_{12} = b_{12}, \ldots, a_{33} = b_{33}$.

The *trace* of a matrix is the sum of the diagonal terms. Thus $\mathrm{tr}(\mathbf{A}) = a_{11} + a_{22} + \cdots + a_{nn}$ for an $n \times n$ matrix. This is only defined for square matrices.

2.3 Operations on matrices

The ordinary arithmetic processes of addition, subtraction, multiplication and division have their counterparts with matrices. With addition and subtraction it is simply a matter of working element by element. For example, if $\mathbf{A}$ and $\mathbf{D}$ are both 3×2 matrices, then

$$\mathbf{A} + \mathbf{D} = \begin{bmatrix} a_{11} & a_{12} \\ a_{21} & a_{22} \\ a_{31} & a_{32} \end{bmatrix} + \begin{bmatrix} d_{11} & d_{12} \\ d_{21} & d_{22} \\ d_{31} & d_{32} \end{bmatrix} = \begin{bmatrix} a_{11} + d_{11} & a_{12} + d_{12} \\ a_{21} + d_{21} & a_{22} + d_{22} \\ a_{31} + d_{31} & a_{32} + d_{32} \end{bmatrix}$$

while

$$\mathbf{A} - \mathbf{D} = \begin{bmatrix} a_{11} - d_{11} & a_{12} - d_{12} \\ a_{21} - d_{21} & a_{22} - d_{22} \\ a_{31} - d_{31} & a_{32} - d_{32} \end{bmatrix}.$$

These operations can only be carried out with two matrices that are of the same size.

In matrix algebra, an ordinary number such as 20 is called a *scalar*. Multiplication of a matrix **A** by a scalar k is then defined as multiplying every element of **A** by k. Thus if **A** is 3×2, as shown above, then

$$k\mathbf{A} = \begin{bmatrix} ka_{11} & ka_{12} \\ ka_{21} & ka_{22} \\ ka_{31} & ka_{32} \end{bmatrix}.$$

Multiplying two matrices together is more complicated. To begin with, $\mathbf{A} \cdot \mathbf{B}$ is only defined if the number of columns of **A** is equal to the number of rows of **B**. If this is the case, say with **A** of size $m \times n$ and **B** of size $n \times p$, then

$$\mathbf{A} \cdot \mathbf{B} = \begin{bmatrix} a_{11} & a_{12} & \cdots & a_{1n} \\ a_{21} & a_{22} & \cdots & a_{2n} \\ \vdots & & & \\ a_{m1} & a_{m2} & \cdots & a_{mn} \end{bmatrix} \cdot \begin{bmatrix} b_{11} & b_{12} & \cdots & b_{1p} \\ b_{21} & b_{22} & \cdots & b_{2p} \\ & & & \\ b_{n1} & b_{n2} & \cdots & b_{np} \end{bmatrix}$$

$$= \begin{bmatrix} \sum a_{1j}b_{j1} & \sum a_{1j}b_{j2} & \cdots & \sum a_{1j}b_{jp} \\ \sum a_{2j}b_{j1} & \sum a_{2j}b_{j2} & \cdots & \sum a_{2j}b_{jp} \\ \vdots & & & \\ \sum a_{mj}b_{j1} & \sum a_{mj}b_{j2} & \cdots & \sum a_{mj}b_{jp} \end{bmatrix},$$

where $\sum$ indicates the summation for all $j = 1, 2, \ldots, n$. Thus the element in the ith row and kth column of $\mathbf{A} \cdot \mathbf{B}$ is

$$\sum a_{ij}b_{jk} = a_{i1}b_{1k} + a_{i2}b_{2k} + \cdots + a_{in}b_{nk}.$$

It is only when **A** and **B** are square that $\mathbf{A} \cdot \mathbf{B}$ and $\mathbf{B} \cdot \mathbf{A}$ are both defined. However, even if this is true, $\mathbf{A} \cdot \mathbf{B}$ and $\mathbf{B} \cdot \mathbf{A}$ are not generally equal. For example,

$$\begin{bmatrix} 2 & -1 \\ 1 & 1 \end{bmatrix} \cdot \begin{bmatrix} 1 & 1 \\ 0 & 1 \end{bmatrix} = \begin{bmatrix} 2 & 1 \\ 1 & 2 \end{bmatrix}$$

while

$$\begin{bmatrix} 1 & 1 \\ 0 & 1 \end{bmatrix} \cdot \begin{bmatrix} 2 & -1 \\ 1 & 1 \end{bmatrix} = \begin{bmatrix} 3 & 0 \\ 1 & 1 \end{bmatrix}.$$

2.4 Matrix inversion

Matrix inversion is analogous to the ordinary arithmetic process of division. Thus for a scalar k it is, of course, true that $k \times k^{-1} = 1$. In a similar way, if $\mathbf{A}$ is a square matrix then its *inverse* is $\mathbf{A}^{-1}$, where $\mathbf{A} \times \mathbf{A}^{-1} = \mathbf{I}$, this being the identity matrix. Inverse matrices are only defined for square matrices. However, all square matrices do not have an inverse. When $\mathbf{A}^{-1}$ exists, it is both a right and left inverse so that $\mathbf{A}^{-1}\mathbf{A} = \mathbf{A}\mathbf{A}^{-1} = \mathbf{I}$.

An example of an inverse matrix is

$$\begin{bmatrix} 2 & 1 \\ 1 & 2 \end{bmatrix}^{-1} = \begin{bmatrix} 2/3 & -1/3 \\ -1/3 & 2/3 \end{bmatrix}.$$

This can be verified by checking that

$$\begin{bmatrix} 2 & 1 \\ 1 & 2 \end{bmatrix} \cdot \begin{bmatrix} 2/3 & -1/3 \\ -1/3 & 2/3 \end{bmatrix} = \begin{bmatrix} 1 & 0 \\ 0 & 1 \end{bmatrix}.$$

Actually, the inverse of a 2×2 matrix, if it exists, can be determined easily. It is given by

$$\begin{bmatrix} a_{11} & a_{12} \\ a_{21} & a_{22} \end{bmatrix}^{-1} = \begin{bmatrix} a_{22}/\Delta & -a_{12}/\Delta \\ -a_{21}/\Delta & a_{11}/\Delta \end{bmatrix},$$

where $\Delta = a_{11}a_{22} - a_{12}a_{21}$. Here the scalar quantity Δ is called the *determinant* of the matrix. Clearly the inverse is not defined if $\Delta = 0$, since finding the elements of the inverse then requires a division by zero. For 3×3 and larger matrices the calculation of an inverse is a tedious business best done using a computer program.

Any square matrix has a determinant value that can be calculated by a generalization of the equation just given for the 2×2 case. If the determinant of a matrix is zero then the inverse does not exist, and vice versa. A matrix with a zero determinant is described as being *singular*.

Matrices sometimes arise for which the inverse is equal to the transpose. They are then described as being *orthogonal*. Thus $\mathbf{A}$ is orthogonal if $\mathbf{A}^{-1} = \mathbf{A}'$.

2.5 Quadratic forms

Let $\mathbf{A}$ be an $n \times n$ matrix and $\mathbf{x}$ be a column vector of length n. Then the expression

$$Q = \mathbf{x}'\mathbf{A}\mathbf{x}$$

is called a *quadratic form*. It is a scalar and can be expressed by the alternative equation

$$Q = \sum_{i=1}^{n} \sum_{j=1}^{n} x_i a_{ij} x_j,$$

where x_i is the element in the ith row of $\mathbf{x}$ and a_{ij} is the element in the ith row and the jth column of $\mathbf{A}$.

2.6 Eigenvalues and eigenvectors

Consider the set of equations

$$a_{11}x_1 + a_{12}x_2 + \cdots + a_{1n}x_n = \lambda x_1$$
$$a_{21}x_1 + a_{22}x_2 + \cdots + a_{2n}x_n = \lambda x_2$$
$$\vdots$$
$$a_{n1}x_1 + a_{n2}x_2 + \cdots + a_{nn}x_n = \lambda x_n$$

which can be written in matrix form as

$$\mathbf{A}\mathbf{x} = \lambda\mathbf{x},$$

or

$$(\mathbf{A} - \lambda\mathbf{I})\mathbf{x} = \mathbf{0},$$

where $\mathbf{I}$ is an $n \times n$ identity matrix and $\mathbf{0}$ is an $n \times 1$ zero vector. It can be shown that these equations can only hold for certain particular values of the scalar λ that are called the *latent roots* or *eigenvalues* of the matrix $\mathbf{A}$. There can be up to n of these roots. Given the ith latent root λ_i, the equations can be solved by arbitrarily setting

$x_1 = 1$. The resulting vector of x values

$$\mathbf{x}_i = \begin{bmatrix} 1 \\ x_{2i} \\ x_{3i} \\ \vdots \\ x_{ni} \end{bmatrix}$$

(or any multiple of it) is called the ith *latent vector* or the ith *eigenvector* of the matrix $\mathbf{A}$. The sum of the eigenvalues of $\mathbf{A}$ is equal to the trace of $\mathbf{A}$.

Finding eigenvalues and eigenvectors is not a simple matter. Like finding a matrix inverse, it is a job best done on a computer.

2.7 Vectors of means and covariance matrices

Populations and samples for a single random variable are often summarized by mean values and variances. Thus suppose that a random sample of size n yields the sample values $x_1, x_2, \ldots, x_n$. Then the *sample mean* is

$$\bar{x} = \sum_{i=1}^{n} x_i/n$$

while the *sample estimate of vatiance* is

$$s^2 = \sum_{i=1}^{n} (x_i - \bar{x})^2/(n-1).$$

These are estimates of the corresponding population parameters – the *population mean* μ, and the *population variance* σ^2.

In a similar way, multivariate populations and samples can be summarized by mean vectors and covariance matrices. These are defined as follows. Suppose that there are p variables $X_1, X_2, \ldots, X_p$ and the values of these for the ith individual in a sample are $x_{i1}, x_{i2}, \ldots, x_{ip}$, respectively. Then the sample mean of variable j is

$$\bar{x}_j = \sum_{i=1}^{n} x_{ij}/n, \tag{2.1}$$

while the sample variance is

$$s_j^2 = \sum_{i=1}^{n} (x_{ij} - \bar{x}_j)^2/(n-1). \qquad (2.2)$$

In addition, the sample *covariance* between variables j and k is defined as

$$c_{jk} = \sum_{i=1}^{n} (x_{ij} - \bar{x}_j)(x_{ik} - \bar{x}_k)/(n-1), \qquad (2.3)$$

this being a measure of the extent to which the two variables are linearly related. The ordinary correlation coefficient for variables j and k, r_{jk}, say, is related to the covariance by the expression

$$r_{jk} = c_{jk}/(s_j s_k). \qquad (2.4)$$

It is clear from equations (2.3) and (2.4) that $c_{kj} = c_{jk}$, $r_{kj} = r_{jk}$ and $r_{kk} = 1$.

The *vector of sample means* is calculated using equation (2.1):

$$\bar{\mathbf{x}} = \begin{bmatrix} \bar{x}_1 \\ \bar{x}_2 \\ \vdots \\ \bar{x}_p \end{bmatrix}. \qquad (2.5)$$

This can be thought of as the centre of the sample. It is an estimate of the *population vector of means*

$$\boldsymbol{\mu} = \begin{bmatrix} \mu_1 \\ \mu_2 \\ \vdots \\ \mu_p \end{bmatrix}. \qquad (2.6)$$

The matrix of variances and covariances

$$\mathbf{C} = \begin{bmatrix} c_{11} & c_{12} & \cdots & c_{1p} \\ c_{21} & c_{22} & \cdots & c_{2p} \\ \vdots & & & \\ c_{p1} & c_{p2} & \cdots & c_{pp} \end{bmatrix}, \qquad (2.7)$$

where $c_{ii} = s_i^2$ is called the *sample covariance matrix*, or sometimes the *sample dispersion matrix*. This reflects the amount of variation in the sample and also the extent to which the p variables are correlated. It is an estimate of the population covariance matrix.

The matrix of correlations as defined by equation (2.4) is

$$
\mathbf{R} = \begin{bmatrix} r_{11} & r_{12} & \cdots & r_{1p} \\ r_{21} & r_{22} & \cdots & r_{2p} \\ \vdots & & & \\ r_{p1} & r_{p2} & \cdots & r_{pp} \end{bmatrix} = \begin{bmatrix} 1 & r_{12} & \cdots & r_{1p} \\ r_{21} & 1 & \cdots & r_{2p} \\ \vdots & \vdots & & \\ r_{p1} & r_{p2} & \cdots & 1 \end{bmatrix}. \tag{2.8}
$$

This is called the *sample correlation matrix*. Like $\mathbf{C}$, this must be symmetric.

2.8 Further reading

A book by Causton (1983) gives a somewhat fuller introduction to matrix algebra than the one given here, but excludes latent roots and vectors. In addition to the chapter on matrix algebra, the other parts of Causton's book provide a good review of general mathematics. A more detailed account of matrix theory, still at an introductory level, is provided by Namboodiri (1984).

These interested in learing more about matrix inversion and finding eigenvectors and eigenvalues, particularly methods for use on microcomputers, will find the book by Nash (1990) a useful source of information.

References

Causton, D.R. (1983) *A Biologist's Basic Mathematics*. Edward Arnold, London.

Namboodiri, K. (1984) *Matrix Algebra: An Introduction*. Sage Publications, Beverly Hills.

Nash, J.C. (1990) *Compact Numerical Methods for Computers*, 2nd edn. Adam Hilger, Bristol.

CHAPTER THREE

Displaying multivariate data

3.1 The problem of displaying many variables in two dimensions

Graphs must be displayed in two dimensions either on paper or on a computer screen. It is therefore straightforward to show one variable plotted on a vertical axis against a second variable plotted on a horizontal axis. For example, Fig. 3.1 shows the alar extent plotted against the total length for the 49 female sparrows measured by Hermon Bumpus in his study of natural selection that has been described in Example 1.1. Such plots allow one or more other characteristics of the objects being studied to be shown as well, and in the case of Bumpus's sparrows survival and non-survival are indicated.

It is considerably more complicated to show one variable plotted against another two, but still possible. For instance, Fig. 3.2 shows beak and head lengths plotted against total lengths and alar lengths for the 49 sparrows. In principle it is also possible to indicate different types of individuals (such as survivors and non-survivors) using different symbols in place of the circles shown here.

It is not possible at all to show one variable plotted against another three at the same time in some extension of a three-dimensional plot. Hence there is a major problem in showing in a simple way the relationships that exist between the individual objects in a multivariate set of data where those objects are each described by four or more variables. Various solutions to this problem have been proposed, and are discussed in this chapter.

3.2 Plotting index variables

One approach to making a graphical summary of the differences between objects that are described by more than four variables

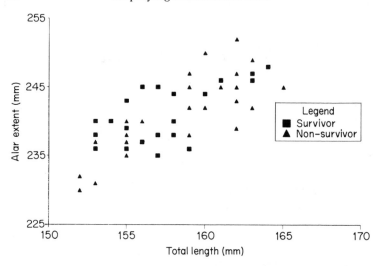

Figure 3.1 Alar extent plotted against total length for the 49 female sparrows measured by Hermon Bumpus.

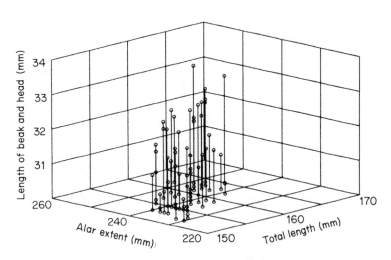

Figure 3.2 Beak and head lengths plotted against total lengths and alar extents for the 49 female sparrows measured by Hermon Bumpus.

involves plotting the objects against the values of two or three index variables. Indeed, a major objective of many multivariate analyses is to produce index variables that can be used for this purpose – a process that is sometimes called ordination. For example, a plot of the values of principal component 2 against the values of principal component 1 can be used as a means of representing the relationships between objects graphically, and a display of principal component 3 against the first two principal components can also be used if necessary.

The use of suitable index variables has the advantage of reducing the problem of plotting many variables to two or three dimensions, but the potential disadvantage that some key difference between the objects may be lost in the reduction. This approach is discussed in various different contexts in the chapters that follow and will not be considered further here.

3.3 The draftsman's display

A draftsman's display of multivariate data consists of a plot of the values for each variable against the values for each of the other variables, with the individual graphs being small enough so that they can all be viewed at the same time. This has the advantage of only needing two-dimensional plots, but the disadvantage that some aspect of the data that is only apparent when three or more variables are considered together will not be apparent.

An example is shown in Fig. 3.3. Here the five variables measured by Hermon Bumpus on 49 sparrows (total length, alar extent, length of beak and head, length of humerus, and length of the keel of the sternum, all in mm) are plotted for the data given in Table 1.1, with an additional first variable being the number of the sparrow, from 1 to 49. Regression lines are also shown on the graphs for each of the individual pairs of variables.

This type of plot is obviously good for showing the relationships between pairs of variables, and highlighting any objects that have unusual values for one or two variables. But because the individual objects are not easily identified it is not immediately clear which objects are similar and which are different. Therefore the plot is not suitable for showing relationships between objects, as distinct from relationships between variables.

Displaying multivariate data

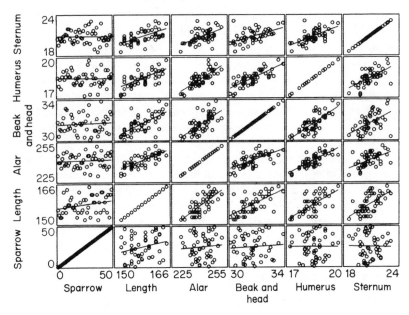

Figure 3.3 Draftsman's display of five variables measured on 49 female sparrows. The line shown in each graph is from the linear regression of the vertical variable on the horizontal variable.

3.4 The symbolic representation of data points

An approach to displaying data that is more truly multivariate involves representing each of the objects for which variables are measured by a symbol, with different characteristics of this symbol varying according to different variables. A number of different types of symbol have been proposed for this purpose including faces (Chernoff, 1973) and stars (Welsch, 1976).

As an illustration, consider the data in Table 1.4 on mean values of six mandible measurements for seven canine groups, as discussed in Example 1.4. Here an important question concerns which of the other groups is most similar to the prehistoric Thai dog, and it can be hoped that this becomes apparent from a graphical comparison of the groups. To this end, Fig. 3.4 shows the data represented by faces and stars.

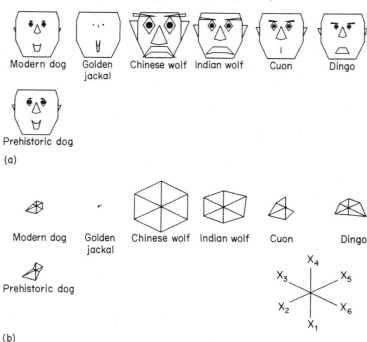

Figure 3.4 Graphical representations of mandible measurements on modern Thai dogs, golden jackals, Chinese wolves, Indian wolves, cuon, dingo and prehistoric Thai dogs: (a) Chernoff faces; (b) stars.

For the faces there was the following connection between features and the variables: mandible breadth to eye size, mandible height to nose size, length of first molar to brow size, breadth of first molar to ear size, length from first to third molar to mouth size, and length from first to fourth premolars to the amount of smile. For example, the eyes are largest for the Chinese wolf with the maximum mandible breadth of 13.5 mm, and smallest for the golden jackal with the minimum mandible length of 8.1 mm. It is apparent from the plots that prehistoric Thai dogs are most similar to modern Thai dogs, and most different from Chinese wolves.

For the stars the six variables were assigned to rays in the order 1, mandible breadth; 2, mandible height; 3, length of first molar; 4, breadth of first molar; 5, length from first to third molar, and 6, length from first to fourth premolars. The mandible length is represented by the ray corresponding to six o'clock and the other variables follow in a clockwise order as indicated with the key that accompanies the figure. Inspection of the stars indicates again that the prehistoric Thai dogs are most similar to modern Thai dogs and most different from Chinese wolves.

Suggestions for alternatives to faces and stars, and a discussion of the relative merits of different symbols, are provided by Everitt (1978) and Toit *et al.* (1986, Chapter 4). In summary it can be said that the use of symbols has the advantage of displaying all variables simultaneously, but the disadvantage that the impression gained from the graph may depend quite strongly on the order in which objects are displayed and the order in which variables are assigned to the different aspects of the symbol.

The assignment of variables is obviously likely to have more effect with faces than it is with stars, because variation in different features of the face may have very different impacts on the observer whereas this is less likely to be the case with different rays of a star. For this reason the recommendation is often made that alternative assignments of variables to features should be tried with faces in order to find the 'best'. The subjective nature of this type of process is clearly unsatisfactory.

3.5 Profiles of variables

Another way to represent objects that are described by several variables that are measured on them is through lines that show the profile of variable values. A simple way to draw these involves just plotting the values for the variables, as shown in Fig. 3.5 for the seven canine groups that have already been considered. The similarity between prehistoric and modern Thai dogs noted from the earlier graphs is still apparent, as is the difference between prehistoric dogs and Chinese wolves. In this graph the variables have been plotted in order of their average values for the seven groups to help in emphasizing similarities and differences.

An alternative representation using bars instead of lines is shown in Fig. 3.6. Here the variables are in their original order because there seems little need to change this when bars are used. The conclusion about similarities and differences between the canine groups would be exactly the same as from Fig. 3.5.

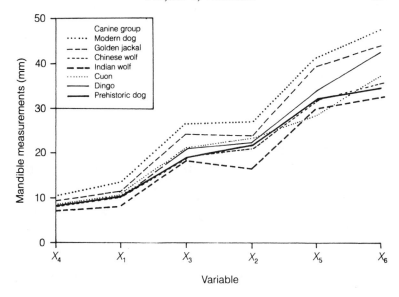

Figure 3.5 Profiles of variables for mandible measurements on modern Thai dogs, golden jackals, Chinese wolves, Indian wolves, cuon, dingo and prehistoric dogs. The variables are in order of increasing average value with X_1 = breadth of mandible, X_2 = height of mandible above first molar, X_3 = length of first molar, X_4 = breadth of first mandible, X_5 = length from first to third molar inclusive, and X_6 = length from first to fourth molar inclusive.

A more complicated type of profile was proposed by Andrews (1972) based on a Fourier series representation. This involves plotting for each object the curve

$$f(t) = x_1/\sqrt{2} + x_2 \sin(t) + x_3 \cos(t) + x_4 \sin(2t)$$
$$+ x_5 \cos(2t) + x_6 \sin(3t) + \ldots, \tag{3.1}$$

for the range of values of t from $-\pi$ to π. Although this appears to be a non-intuitive method for representing the data it does have several useful properties such as in a sense preserving the 'distance' between objects. See Andrews's paper for more details. The method is mentioned here because it has received considerable use, and is often claimed to be better than faces and stars.

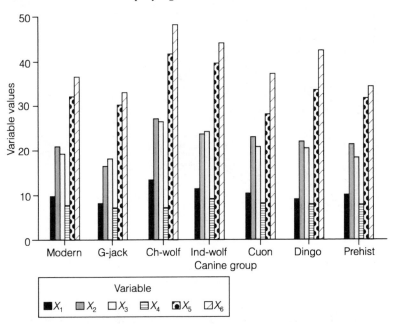

Figure 3.6 An alternative way to show profiles using bars instead of the lines used in Fig. 3.5. There are obvious abbreviations for the names of the groups (modern Thai dogs, golden jackals, Chinese wolves, Indian wolves, cuon, dingo and prehistoric Thai dogs).

For the canine data the Andrews plots are shown in Fig. 3.7. Each of the six variables was standardized by subtracting the overall mean and dividing by the overall standard deviation before the Fourier functions were determined from equation (3.1). For example, all of the values for the first variable (the breadth of the mandible) were coded to (breadth − 10.486)/1.572 before the function (3.1) was calculated, where 10.486 is the mean and 1.572 is the standard deviation of the original variable over the seven groups. This coding serves two purposes. First, it ensures that each of the variables contributes equally to the variation in the functions. Second, it eliminates variation in the functions due to the variables having non-zero means.

Inspection of Fig. 3.7 shows relationships between the seven groups that are similar to those shown by the other plots already considered. In particular, prehistoric Thai dogs are seen to resemble modern Thai dogs most closely and differ most from Chinese wolves.

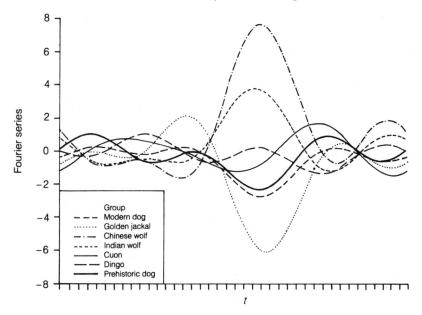

Figure 3.7 Andrew's (1972) Fourier series plot using equation (3.1) for the six variables measured on seven canine groups. The plots were calculated with the variables X_1 to X_6 as defined in Fig. 3.5 but with each variable standardized to have a mean of zero and a standard deviation of one over the seven groups.

3.6 Discussion and further reading

It seems fair to say that there is no method for displaying data on many variables at a time that is completely satisfactory if it is not desirable to reduce these variables to two or three index variables (using one of the methods discussed in the remainder of this book). The three types of method that have been discussed here involve the use of a draftsman's display with all pairs of variables plotted against each other, symbols (stars or faces), and lines (variable profiles or Fourier series plots). Which of these is most suitable for a particular application depends on the circumstances, but as a general rule the draftsman's display highlights relationships between pairs of variables, the use of symbols highlights unusual cases, and the use of lines is good for indicating clusters of cases. If simplicity is a key consideration then the draftsman's display, star plots and variable profiles seem to have much to recommend them.

For further information about the theory of the construction of graphs in general see the books by Cleveland (1985) and Tufte (1983, 1990). For more details on graphical methods for multivariate data see the books by Everitt (1978) and Toit *et al.* (1986).

Figures 3.1 and 3.5–3.7 were originally produced using Word-Perfect Presentations (WordPerfect, 1992), and Figures 3.2–3.4 were originally produced using SOLO (BMDP, 1989).

References

Andrews, D.F. (1972) Plots of high-dimensional data. *Biometrics* **28**, 125–36.

BMDP (1989) *SOLO User's Guide, Version 3.1.* BMDP Statistical Software Inc., 1440 Sepulveda Boulevard, Los Angeles, California 90025.

Chernoff, H. (1973) Using faces to represent points in *K*-dimensional space graphically. *Journal of the American Statistical Association* **68**, 361–8.

Cleveland, W.S. (1985) *The Elements of Graphing Data.* Wadsworth, Monterey, California.

Everitt, B. (1978) *Graphical Techniques for Multivariate Data.* Heinemann, London.

Toit, S.H.C., Steyn, A.G.W. and Stumf, R.H. (1986) *Graphical Exploratory Data Analysis.* Springer-Verlag, New York.

Tufte, E.R. (1983) *The Visual Display of Quantitative Information.* Graphics Press, Cheshire, Connecticut.

Tufte, E.R. (1990) *Envisioning Information.* Graphics Press, Cheshire, Connecticut.

Welsch, R.E. (1976) Graphics for data analysis. *Computers and Graphics* **2**, 31–7.

WordPerfect (1992) *WordPerfect Presentations, Version 2.0.* WordPerfect Corporation, 1555 North Technology Way, Orem, Utah 84057.

CHAPTER FOUR

Tests of significance
with multivariate data

4.1 Simultaneous tests on several variables

When data are collected for several variables on the same sample units then it is always possible to examine the variables one at a time as far as tests of significance are concerned. For example, if the sample units are in two groups then a difference between the means for the two groups can be tested separately for each variable. Unfortunately, there is a drawback to this approach because of the repeated use of significance tests, each of which has a certain probability of leading to a wrong conclusion. Basically, as will be discussed further in section 4.4, the probability of falsely finding at least one significant difference accumulates with the number of tests carried out so that it may become unacceptably large.

There are ways of adjusting significance levels in order to allow for multiple testing but it may be preferable to conduct a single test that uses the information from all variables together. For example, it might be desirable to test the hypothesis that the means of all variables are the same for two multivariate populations, with a significant result being taken as evidence that the means differ for at least one variable. These types of overall test are considered in this chapter for the comparison of means and the comparison of variation, for two or more samples.

4.2 Comparison of mean values for two samples: single variable case

Consider the data in Table 1.1 on the body measurements of 49 female sparrows. Consider in particular the first measurement, which is total length. A question of some interest might be whether the mean of this variable was the same for survivors and non-survivors of the storm that led to the birds being collected. There is then a sample (hopefully random) of 21 survivors and a second sample (again hopefully random) of 28 non-survivors. We wish to know whether the two sample means are significantly different. A standard approach would be to carry out a t test.

Thus, suppose that in a general situation there is a single variable X and two random samples of values are available from different populations. Let x_{i1} denote the values of X in the first sample, for $i = 1, 2, \ldots, n_1$, and x_{i2} denote the values in the second sample, for $i = 1, 2, \ldots, n_2$. Then the mean and variance for the jth sample are

$$\bar{x}_j = \sum_{i=1}^{n_j} x_{ij}/n_j$$

and

$$s_j^2 = \sum_{i=1}^{n} (x_{ij} - \bar{x}_j)^2/(n_j - 1). \tag{4.1}$$

On the assumption that X is normally distributed in both samples, with a common within-sample variance, a test to see whether the two sample means are significantly different involves calculating the statistic

$$t = (\bar{x}_1 - \bar{x}_2) \Big/ \left\{ s \sqrt{\left(\frac{1}{n_1} + \frac{1}{n_2} \right)} \right\} \tag{4.2}$$

and seeing whether this is significantly different from zero in comparison with the t distribution with $n_1 + n_2 - 2$ degrees of freedom (d.f.). Here

$$s^2 = \{(n_1 - 1)s_1^2 + (n_2 - 1)s_2^2\}/(n_1 + n_2 - 2) \tag{4.3}$$

is the pooled estimate of variance from the two samples.

It is known that this test is fairly robust to the assumption of normality. Providing that the population distributions of X are not too different from normal it should be satisfactory, particularly for sample sizes of about 20 or more. The assumption of equal within-sample variances is also not too crucial. Providing that the ratio of the true variances is within the limits 0.4 to 2.5, inequality of variance will have little adverse effect on the test. The test is particularly robust if the two sample sizes are equal, or nearly so (Carter *et al.*, 1979). If the population variances are very different then the t test can be modified to allow for this (Dixon and Massey, 1969, p. 119).

4.3 Comparison of mean values for two samples: multivariate case

Consider again the sparrow data of Table 1.1. The test described in the previous section can obviously be employed for each of the five measurements shown in the table (total length, alar extent, length of beak and head, length of humerus, and length of keel of sternum). In that way it is possible to decide which, if any, of these variables appear to have had different mean values for survivors and non-survivors. However, in addition to these it may also be of some interest to know whether all five variables considered together suggest a difference between survivors and non-survivors. In other words: does the total evidence point to mean differences between survivors and non-survivors?

What is needed to answer this question is a multivariate test. One possibility is Hotelling's T^2 test. The statistic used is then a generalization of the t statistic of equation (4.2) or, to be more precise, the square of the t statistic.

In a general case there will be p variables $X_1, X_2, \ldots, X_p$ being considered, and two samples with sizes n_1 and n_2. There are then two-sample mean vectors $\mathbf{x}_1$ and $\mathbf{x}_2$, with each one being calculated as shown in equations (2.1) and (2.5). There are also two-sample covariance matrices, $\mathbf{C}_1$ and $\mathbf{C}_2$, with each one being calculated as shown in equations (2.2), (2.3) and (2.7).

Assuming that the population covariance matrices are the same for both populations, a pooled estimate of this matrix is

$$\mathbf{C} = \{(n_1 - 1)\mathbf{C}_1 + (n_2 - 1)\mathbf{C}_2\}/(n_1 + n_2 - 2). \tag{4.4}$$

Hotelling's T^2 statistic is defined as

$$T^2 = n_1 n_2 (\bar{\mathbf{x}}_1 - \bar{\mathbf{x}}_2)' \mathbf{C}^{-1} (\bar{\mathbf{x}}_1 - \bar{\mathbf{x}}_2)/(n_1 + n_2). \tag{4.5}$$

A significantly large value for this statistic is evidence that the mean vectors are different for the two sampled populations. The significance or lack of significance of T^2 is most simply determined by using the fact that in the null hypothesis case of equal population means the transformed statistic

$$F = (n_1 + n_2 - p - 1)T^2/\{(n_1 + n_2 - 2)p\} \qquad (4.6)$$

follows an F distribution with p and $(n_1 + n_2 - p - 1)$ d.f.

Because T^2 is a quadratic form it is a scalar which can be written in the alternative way

$$T^2 = \frac{n_1 n_2}{n_1 + n_2} \sum_{i=1}^{p} \sum_{k=1}^{p} (\bar{x}_{1i} - \bar{x}_{2i})c^{ik}(\bar{x}_{1k} - \bar{x}_{2k}), \qquad (4.7)$$

which may be simpler to compute. Here $\bar{x}_{jl}$ is the mean of variable X_l in the jth sample and c^{ik} is the element in the ith row and kth column of the inverse matrix $\mathbf{C}^{-1}$.

Hotelling's T^2 statistic is based on an assumption of normality and equal within-sample variability. To be precise, the two samples being compared using the T^2 statistic are assumed to come from multivariate normal distributions with equal covariance matrices. Some deviation from multivariate normality is probably not serious. A moderate difference between population covariance matrices is also not too important, particularly with equal or nearly equal sample sizes (Carter *et al.*, 1979). If the two population covariance matrices are very different, and sample sizes are very different as well, then a modified test can be used (Yao, 1965).

Example 4.1 Testing mean values for Bumpus's female sparrows

As an example of the use of the univariate and multivariate tests that have been described for two samples, consider the sparrow data shown in Table 1.1. Here it is a question of whether there is any difference between survivors and non-survivors with respect to the mean values of five morphological characters.

First of all, tests on the individual variables can be considered, starting with X_1, the total length. The mean of this variable for the 21 survivors is $\bar{x}_1 = 157.38$ while the mean for the 28 non-survivors is $\bar{x}_2 = 158.43$. The corresponding sample variances are $s_1^2 = 11.05$

and $s_2^2 = 15.07$. The pooled variance from equation (4.3) is therefore

$$s^2 = (20 \times 11.05 + 27 \times 15.07)/47 = 13.36,$$

and the t statistic of equation (4.2) is

$$t = (157.38 - 158.43) \Big/ \sqrt{\left\{ 13.36 \left(\frac{1}{21} + \frac{1}{28} \right) \right\}} = -0.99,$$

with $n_1 + n_2 - 2 = 47$ d.f. This is not significantly different from zero at the 5% level so there is no evidence of a mean difference between survivors and non-survivors with regard to total length.

Table 4.1 summarizes the results of tests on all five of the variables in Table 1.1 taken individually. In no case is there any evidence of a mean difference between survivors and non-survivors.

For tests on all five variables considered together it is necessary to know the sample mean vectors and covariance matrices. The means are given in Table 4.1. The covariance matrices are defined by equation (2.7). For the sample of 21 survivors,

$$\bar{\mathbf{x}}_1 = \begin{bmatrix} 157.381 \\ 241.000 \\ 31.433 \\ 18.500 \\ 20.810 \end{bmatrix} \text{ and } \mathbf{C}_1 = \begin{bmatrix} 11.048 & 9.100 & 1.557 & 0.870 & 1.286 \\ 9.100 & 17.500 & 1.910 & 1.310 & 0.880 \\ 1.557 & 1.910 & 0.531 & 0.189 & 0.240 \\ 0.870 & 1.310 & 0.189 & 0.176 & 0.133 \\ 1.286 & 0.880 & 0.240 & 0.133 & 0.575 \end{bmatrix}.$$

Table 4.1 Comparison of mean values for survivors and non-survivors for Bumpus's female sparrows with variables taken one at a time

| Variable | Survivors | | Non-survivors | | |
	$\bar{x}_1$	s_1^2	$\bar{x}_2$	s_2^2	t (47.d.f.)
Total length	157.38	11.05	158.43	15.07	−0.99
Alar extent	241.00	17.50	241.57	32.55	−0.39
Length beak & head	31.43	0.53	31.48	0.73	−0.20
Length humerus	18.50	0.18	18.45	0.43	0.33
Length keel of sternum	20.81	0.58	20.84	1.32	−0.10

For the sample of 28 non-survivors,

$$
\bar{\mathbf{x}}_2 = \begin{bmatrix} 158.429 \\ 241.571 \\ 31.479 \\ 18.446 \\ 20.839 \end{bmatrix} \text{ and } \mathbf{C}_2 = \begin{bmatrix} 15.069 & 17.190 & 2.243 & 1.746 & 2.931 \\ 17.190 & 32.550 & 3.398 & 2.950 & 4.066 \\ 2.243 & 3.398 & 0.728 & 0.470 & 0.559 \\ 1.743 & 2.950 & 0.470 & 0.434 & 0.506 \\ 2.931 & 4.066 & 0.559 & 0.506 & 1.321 \end{bmatrix}.
$$

The pooled sample covariance matrix is then

$$
\mathbf{C} = (20\mathbf{C}_1 + 27\mathbf{C}_2)/47 = \begin{bmatrix} 13.358 & 13.748 & 1.951 & 1.373 & 2.231 \\ 13.748 & 26.146 & 2.765 & 2.252 & 2.710 \\ 1.951 & 2.765 & 0.645 & 0.350 & 0.423 \\ 1.373 & 2.252 & 0.350 & 0.324 & 0.347 \\ 2.231 & 2.710 & 0.423 & 0.347 & 1.004 \end{bmatrix}
$$

where, for example, the element in the second row and third column is $(20 \times 1.910 + 27 \times 3.398)/47 = 2.765$.

The inverse of the matrix $\mathbf{C}$ is found to be

$$
\mathbf{C}^{-1} = \begin{bmatrix} 0.2061 & -0.0694 & -0.2395 & 0.0785 & -0.1969 \\ -0.0694 & 0.1234 & -0.0376 & -0.5517 & 0.0277 \\ -0.2395 & -0.0376 & 4.2219 & -3.2624 & -0.0181 \\ 0.0785 & -0.5517 & -3.2624 & 11.4610 & -1.2720 \\ -0.1969 & 0.0277 & -0.0181 & -1.2720 & 1.8068 \end{bmatrix}.
$$

This can be verified by evaluating the product $\mathbf{C} \times \mathbf{C}^{-1}$ and seeing that this is a unit matrix (apart from rounding errors).

Substituting the elements of $\mathbf{C}^{-1}$ and other values into equation (4.7) produces

$$
\begin{aligned}
T^2 = {} & \frac{21 \times 28}{21 + 28}[(157.381 - 158.429) \times 0.2061 \times (157.381 - 158.429) \\
& - (157.318 - 158.429) \times 0.0694 \times (241.000 - 241.571) \\
& + \cdots + (20.810 - 20.839) \times 1.8068 \times (20.810 - 20.839)] \\
= {} & 2.824.
\end{aligned}
$$

Using equation (4.6) this converts to an F statistic of

$$F = (21 + 28 - 5 - 1) \times 2.824 / \{(21 + 28 - 2) \times 5\} = 0.517,$$

with 5 and 43 d.f. Clearly this is not significantly large because a significant F value must exceed unity. Hence there is no evidence of a difference in means for survivors and non-survivors, taking all five variables together.

4.4 Multivariate versus univariate tests

In this last example there were no significant results either for the variables considered individaully or for the overall multivariate test. It should be noted, however, that it is quite possible to have insignificant univariate tests but a significant multivariate test. This can occur because of the accumulation of the evidence from the individual variables in the overall test. Conversely, an insignificant multivariate test can occur when some univariate tests are significant because the evidence of a difference provided by the significant variables is swamped by the evidence of no difference provided by the other variables.

One important aspect of the use of a multivariate test as distinct from a series of univariate tests concerns the control of type one error rates. A *type one error* involves finding a significant result when, in reality, the two samples being compared come from the same population. With a univariate test at the 5% level there is a 0.95 probability of a non-significant result when the population means are the same. Hence if p independent tests are carried out under these conditions then the probability of getting no significant results is 0.96^p. The probability of at least one significant result is therefore $1 - 0.95^p$. With many tests this can be quite a large probability. For example, if p is 5, the probability of at least one significant result by chance alone is $1 - 0.95^5 = 0.23$. With multivariate data, variables are usually not independent so $1 - 0.95^p$ does not quite give the correct probability of at least one significant result by chance alone if variables are tested one by one with univariate t tests. However, the principle still applies: the more tests that are made, the higher the probability of obtaining at least one significant result by chance.

On the other hand, a multivariate test such as Hotelling's T^2 test using the 5% level of significance gives a 0.05 probability of a type one error, irrespective of the number of variables involved. This is a distinct advantage over a series of univariate tests, particularly when the number of variables is large.

There are ways of adjusting significance levels in order to control the overall probability of a type one error when several tests are carried out. However, the use of a single multivariate test provides a better alternative procedure in many cases. A multivariate test has the added advantage of taking proper account of the correlation between variables.

4.5 Comparison of variation for two samples: single variable case

With a single variable, the best known method for comparing the variation in two samples is the F test. If s_j^2 is the variance in the jth sample, calculated as shown in equation (4.1), then the ratio s_1^2/s_2^2 is compared with percentage points of the F distribution with $(n_1 - 1)$ and $(n_2 - 1)$ d.f. Unfortunately, the F test is known to be rather sensitive to the assumption of normality. A significant result may well be due to the fact that a variable is not normally distributed rather than to unequal variances. For this reason it is sometimes argued that the F test should never be used to compare variances.

A robust alternative to the F test is Levene's (1960) test. The idea here is to transform the original data into absolute deviations from the mean and then test for a significant difference between the mean deviations in the two samples, using a t test. Absolute deviations from the arithmetic mean are usually used but a more robust test is possible by using absolute deviations from sample medians (Schultz, 1983). The procedure is illustrated in Example 4.2 below.

4.6 Comparison of variation for two samples: multivariate case

Most textbooks on multivariate methods suggest the use of Bartlett's test to compare the variation in two multivariate samples. This is described, for example, by Srivastava and Carter (1983, p. 333). However, this test is rather sensitive to the assumption that the samples are from multivariate normal distributions. There is always the possibility that a significant result is due to non-normality rather than to unequal population covariance matrices.

An alternative procedure that should be more robust can be constructed using the principle behind Levene's test. Thus the data values can be transformed into absolute deviations from sample means or medians. The question of whether two samples display significantly different amounts of variation is then transformed into a question of whether the transformed values show significantly different mean vectors. Testing of the mean vectors can be done using a T^2 test.

Another possibility was suggested by Van Valen (1978). This involves calculating

$$d_{ij} = \sqrt{\left\{ \sum_{k=1}^{p} (x_{ijk} - \bar{x}_{jk})^2 \right\}}, \qquad (4.8)$$

where x_{ijk} is the value of variable X_k for the ith individual in sample j, and $\bar{x}_{jk}$ is the mean of the same variable in the sample. The sample means of the d_{ij} values are compared by a t test. Obviously if one sample is more variable than another then the mean d_{ij} value will be higher in the more variable sample.

To ensure that all variables are given equal weight, they should be standardized before the calculation of the d_{ij} values. Coding them to have unit variances will achieve this. For a more robust test it may be better to use sample medians in place of the sample means in equation (4.8). Then the formula for d_{ij} values is

$$d_{ij} = \sqrt{\left\{ \sum_{k=1}^{p} (x_{ijk} - M_{jk})^2 \right\}}, \qquad (4.9)$$

where M_{jk} is the median for variable X_k in the jth sample.

The T^2 test and Van Valen's test for deviations from medians are illustrated in the example that follows.

One point to note about the use of the test statistics (4.8) and (4.9) is that they are based on an implicit assumption that if the two samples being tested differ, then one sample will be more variable than the other for all variables. A significant result cannot be expected in a case where, for example, X_1 and X_2 are more variable in sample 1 but X_3 and X_4 are more variable in sample 2. The effect of the differing variances would then tend to cancel out in the calculation of d_{ij}. Thus Van Valen's test is not appropriate for situations where changes in the level of variation are not expected to be consistent for all variables.

Example 4.2 Testing variation for Bumpus's female sparrows

With Bumpus's data shown in Table 1.1, the most interesting question concerns whether the non-survivors were more variable than the survivors. This is what is expected if stabilizing selection took place.

First of all, the individual variables can be considered one at a time, starting with X_1, the total length. For Levene's test the original data values are transformed into deviations from sample medians. The median for survivors is 157 mm. The absolute deviations from this for the 21 birds in the sample then have a mean of $\bar{x}_1 = 2.57$ and a variance of $s_1^2 = 4.26$. The median for non-survivors is 159 mm. The absolute deviations from this median for the 28 birds in the sample have a mean of $\bar{x}_2 = 3.29$ with a variance of $s_2^2 = 4.21$. The pooled variance from equation (4.3) is 4.231 and the t statistic of equation (4.2) is

$$t = (2.57 - 3.29) \bigg/ \sqrt{\left[4.231 \left\{ \frac{1}{21} + \frac{1}{28} \right\} \right]} = -1.21,$$

with 47 d.f.

Because non-survivors would be more variable than survivors if stabilizing selection occurred, it is a one-sided test that is required here, with low values of t providing evidence of selection. Clearly the observed value of t is not significantly low in the present instance. The t values for the other variables are as follows: alar extent, $t = -1.18$; length of beak and head, $t = -0.81$; length of humerus, $t = -1.91$; length of keel of sternum, $t = -1.40$. Only for the length of humerus is the result significantly low at the 5% level.

Table 4.2 shows absolute deviations from sample medians for the data after they have been standardized. For example, the first value given for variable 1, for survivors, is 0.28. This was obtained as follows. First, the original data were coded to have a zero mean and a unit variance for all 49 birds. This transformed the total length for the first survivor to $(156 - 157.98)/3.617 = -0.55$. The median transformed length for survivors was then -0.27. Hence the

Table 4.2 Absolute deviations from sample medians for Bumpus's data and *d* values from equation (4.9)

Bird	Total length	Alar extent	Length beak & head	Length humerus	Length keel sternum	*d*
1	0.28	1.00	0.25	0.00	0.10	1.07
2	0.83	0.00	1.27	1.07	1.02	2.12
3	1.11	0.00	0.51	0.18	0.00	1.23
4	1.11	0.80	0.64	1.43	0.41	2.12
5	0.55	0.60	0.13	0.18	0.31	0.90
6	1.66	1.40	0.76	0.90	0.31	2.49
7	0.00	0.40	0.64	0.18	0.41	0.87
8	0.55	0.20	1.78	0.18	0.61	1.98
9	1.94	1.59	1.65	1.07	0.51	3.23
10	0.28	0.40	0.51	0.54	1.43	1.68
11	0.28	0.00	0.13	0.18	1.43	1.47
12	0.83	0.80	0.38	0.18	0.10	1.23
13	1.11	1.20	1.14	1.43	1.22	2.74
14	0.00	1.00	0.76	1.07	0.61	1.76
15	0.00	1.00	0.13	0.72	0.82	1.48
16	0.28	0.60	0.64	0.90	0.31	1.32
17	0.28	0.80	0.00	0.00	1.02	1.32
18	1.11	0.40	1.14	0.54	0.31	1.75
19	0.55	0.80	1.40	0.00	0.51	1.78
20	1.66	1.20	1.40	0.18	1.32	2.82
21	0.55	0.80	0.13	0.90	0.92	1.61
22	1.11	0.40	0.13	0.90	0.00	1.48
23	0.83	0.40	0.00	0.54	0.10	1.07
24	0.28	0.00	1.40	0.54	1.02	1.83
25	1.94	1.99	1.53	2.33	0.92	4.04
26	0.28	1.59	0.25	0.54	1.83	2.52
27	1.11	1.00	0.64	0.00	0.71	1.77
28	0.55	0.60	0.89	1.79	0.71	2.27
29	1.66	0.60	2.03	2.33	2.04	4.10
30	1.66	2.19	1.78	2.15	0.92	4.02
31	0.83	0.60	1.53	0.90	2.45	3.19
32	0.83	0.20	0.13	0.54	0.61	1.19
33	0.00	0.60	0.38	0.00	1.02	1.24
34	0.00	1.00	0.76	0.72	1.73	2.26
35	1.11	0.20	0.76	0.00	0.61	1.49
36	0.83	1.99	0.51	1.07	1.53	2.90
37	1.94	2.39	1.40	2.15	2.14	4.54
38	0.00	0.00	0.89	0.54	0.20	1.06
39	1.11	0.80	0.38	1.07	1.43	2.28
40	1.11	1.40	2.42	1.79	2.14	4.10
41	1.11	0.00	0.64	0.72	0.00	1.46

Table 4.2 (*Contd.*)

Bird	Total length	Alar extent	Length beak & head	Length humerus	Length keel sternum	d
42	0.83	1.00	0.25	0.54	0.41	1.48
43	0.00	0.80	0.00	0.18	0.41	0.91
44	0.55	0.60	0.76	1.07	0.10	1.55
45	1.11	1.40	1.02	1.43	1.12	2.74
46	0.83	1.00	0.51	1.07	0.31	1.79
47	1.66	1.00	1.14	0.18	0.31	2.28
48	0.83	0.60	1.27	0.00	0.41	1.68
49	1.38	1.20	1.02	0.54	0.20	2.17

absolute deviation from the sample median for the first survivor is $|-0.55 - (-0.27)| = 0.28$, as recorded.

Comparing the transformed sample mean vectors for the five variables using Hotelling's T^2 test gives a test statistic of $T^2 = 4.75$, corresponding to an F statistic of 0.87 with 5 and 43 d.f. (equations (4.7) and (4.6)). There is therefore no evidence of a significant difference between the samples from this test.

Finally, consider Van Valen's test. The d values from equation (4.9) are shown in the last column of Table 4.2. The mean for survivors is 1.760, with variance 0.411. The mean for non-survivors is 2.265, with variance 1.133. The t value from equation (4.2) is then -1.92, which is significantly low at the 5% level. Hence this test indicates more variation for non-survivors than for survivors.

An explanation for the significant result with this test, but no significant result with the T^2 test, is not hard to find. As noted above, the T^2 test is not directional. Thus if the first sample has large means for some variables and small means for others when compared to the second sample, then all of the differences contribute to T^2. On the other hand, Van Valen's test is specifically for less variation in sample 1 than in sample 2, for all variables. In the present case all of the variables show less variation in sample 1 than in sample 2. Van Valen's test has emphasized this fact but the T^2 test has not.

4.7 Comparison of means for several samples

When there is a single variable and several samples to be compared, the generalization of the t test is the F test from a one-factor analysis of variance. The calculations are as shown in Table 4.3.

When there are several variables and several samples, a so-called 'likelihood ratio test' can be used to compare the sample mean vectors. This involves calculating the statistic

$$\phi = [n - 1 - \tfrac{1}{2}(p + m)]\log_e[|\mathbf{T}|/|\mathbf{W}|] \tag{4.10}$$

where n is the total number of observations, p is the number of variables, m is the number of samples, $|\mathbf{T}|$ is the determinant of the total sum of squares and cross-products matrix, and $|\mathbf{W}|$ is the determinant of the within-sample sum of squares and cross-products matrix. This statistic can be tested for significance by comparison with the chi-squared distribution with $p(m - 1)$ d.f. In Example 4.3 below the likelihood ratio test is used to compare the means for the five samples of male Egyptian skulls provided in Table 1.2.

The matrices $\mathbf{T}$ and $\mathbf{W}$ require some further explanation. Let x_{ijk} denote the value of variable X_k for the ith individual in the jth sample, $\bar{x}_{jk}$ denote the mean of X_k in the same sample, and $\bar{x}_k$ denote the overall mean of X_k for all the data taken together. Then the element in row r and column c of $\mathbf{T}$ is

$$t_{rc} = \sum_{j=1}^{m} \sum_{i=1}^{n_j} (x_{ijr} - \bar{x}_r)(x_{ijc} - \bar{x}_c). \tag{4.11}$$

The element in the rth row and cth column of $\mathbf{W}$ is

$$w_{rc} = \sum_{j=1}^{m} \sum_{i=1}^{n_j} (x_{ijr} - \bar{x}_{jr})(x_{ijc} - \bar{x}_{jc}). \tag{4.12}$$

The test based on equation (4.10) involves the assumption that the distribution of the p variables is multivariate normal, with a constant within-sample covariance matrix. It is probably a fairly robust test in the sense that moderate deviations from this assumption do not unduly affect the characteristics of the test.

Table 4.3 One-factor analysis of variance for a single variable and m samples

Source of variation	Sum of squares	Degrees of freedom	Mean square	F
Between samples	$B = T - W$	$m - 1$	$M_1 = B/(m-1)$	M_1/M_2
Within samples	$W = \sum_{j=1}^{m} \sum_{i=1}^{n_j} (x_{ij} - \bar{x}_j)^2$	$n - m$	$M_2 = W/(n-m)$	
Total	$T = \sum_{j=1}^{m} \sum_{i=1}^{n_j} (x_{ij} - \bar{x})^2$	$n - 1$		

n_j = size of jth sample

$n = \sum_{j=1}^{m} n_j$ = total number of observations

x_{ij} = ith observation in jth sample

$\bar{x}_j = \sum_{i=1}^{n_j} x_{ij}/n_j$ = mean of jth sample

$\bar{x} = \sum_{j=1}^{m} \sum_{i=1}^{n_j} x_{ij}/n$ = overall mean

4.8 Comparison of variation for several samples

Bartlett's test is the best known for comparing the variation in several samples. This test has already been mentioned for the two-sample situation with several variables to be compared. See Srivastava and Carter (1983, p. 333) for details of the calculations involved. The test can be used with one or several variables. However, it does have the problem of being rather sensitive to deviations from normality in the distribution of the variables being considered.

Here, robust alternatives to Bartlett's test are recommended, these being generalizations of what was suggested for the two-sample situation. Thus absolute deviations from sample medians can be calculated for the data in m samples. For a single variable these can be treated as the observations for a one-factor analysis of variance. A significant F ratio is then evidence that the samples come from populations with different mean deviations, i.e., populations with different variability. If the ϕ statistic of equation (4.10) is calculated from the transformed data for p variables, then a significant result indicates that the covariance matrix is not constant for the m populations sampled.

Alternatively, the variables can be standardized to have unit variances for all the data lumped together and d values calculated using equation (4.9). These can then be analysed by a one-factor analysis of variance. This generalizes Van Valen's test that was suggested for comparing the variation in two multivariate samples. A significant F ratio from the analysis of variance indicates that some of the m populations sampled are more variable than others. As in the two-sample situation, this test is only really appropriate when some samples may be more variable than others for all the measurements being considered.

Example 4.3 Comparison of samples of Egyptian skulls

As an example of the test for comparing several samples, consider the data shown in Table 1.2 for four measurements on male Egyptian skulls for five samples of different ages.

A one-factor analysis of variance on the first variable, maximum breadth, provides $F = 5.95$, with 4 and 145 d.f. (Table 4.3). This is significantly large at the 0.1% level and hence there is clear evidence that the mean changed with time. For the other three variables, analysis of variance provides the following results: basibregmatic height, $F = 2.45$ (significant at the 5% level); basialveolar length $F = 8.31$ (significant at the 0.1% level); nasal height, $F = 1.51$ (not significant). It appears that the mean changed with time for the first three variables.

Next, consider the four variables together. If the five samples are combined then the matrix of sums of squares and products for the 150 observations, calculated using equation (4.11), is

$$\mathbf{T} = \begin{bmatrix} 3563.89 & -222.81 & -615.16 & 426.73 \\ -222.81 & 3635.17 & 1046.28 & 346.47 \\ -615.16 & 1046.28 & 4309.27 & -16.40 \\ 426.73 & 346.47 & -16.40 & 1533.33 \end{bmatrix},$$

for which the determinant is $|\mathbf{T}| = 7.306 \times 10^{13}$. The within-sample matrix of sums of squares and cross-products is found from equation (4.12) to be

$$\mathbf{W} = \begin{bmatrix} 3061.07 & 5.33 & 11.47 & 291.30 \\ 5.33 & 3405.27 & 754.00 & 412.53 \\ 11.47 & 754.00 & 3505.97 & 164.33 \\ 291.30 & 412.53 & 164.33 & 1472.13 \end{bmatrix},$$

for which the determinant is $|\mathbf{W}| = 4.848 \times 10^{13}$. Substituting $n = 150$, $p = 4$, $m = 5$ and the values of $|\mathbf{T}|$ and $|\mathbf{W}|$ into equation (4.10) then yields $\phi = 61.31$, with $p(m - 1) = 16$ d.f. This is significantly large at the 0.1% level in comparison with the chi-squared distribution. There is therefore clear evidence that the vector of mean values of the four variables changed with time.

For comparing the amount of variation in the samples it is a straightforward matter to transform the data into absolute deviations from sample medians. Analysis of variance then shows no significant difference between the sample means of the transformed data for any of the four variables. The ϕ statistic is not significant for all variables taken together. Also, analysis of variance shows no significant difference between the mean d values calculated using equation (4.9).

It appears that mean values changed with time for the four variables being considered but the variation about the means remained fairly constant.

4.9 Computational methods and computer programs

The multivariate tests discussed in this chapter are not all readily available in standard statistical computer programs so a few remarks are in order about methods for carrying them out.

Hotelling's T^2 test (section 4.3) is often available in standard programs, and if necessary can be carried out using one of the readily available spreadsheet programs that permits matrix inversion, such as LOTUS 123 (Lotus Development Corporation, 1992). With this approach the most difficult part may in fact be the calculation of the sample covariance matrices. These comments also apply with the use of the T^2 test on deviations from sample means or medians to compare the amount of variation in two samples (section 4.6).

Van Valen's test (also in section 4.6) is probably carried out most easily in a spreadsheet program because it does not require the calculation of covariance matrices or need matrix operations.

The likelihood ratio test to compare the mean vectors of several samples (section 4.7) is one of the most readily available multivariate tests and there should be little difficulty in finding a program to carry this out. The requirement to calculate determinants of matrices means that most spreadsheet programs are not suitable for this purpose.

Bartlett's test for comparing the variation in several samples (mentioned in section 4.8) is provided in many standard programs. The other tests for the same purpose that have been discussed briefly are probably best carried out by converting the observations to deviations from means or medians using a spreadsheet program or some other convenient tool and then carrying out the remaining calculations with a program for the likelihood ratio test or a one-factor analysis of variance.

Exercise

Example 1.4 concerned the comparison between prehistoric dogs from Thailand and six other related animal groups in terms of mean mandible measurements. Table 4.4 shows some further data for the comparison of these groups that are part of the more extensive data discussed in the paper by Higham *et al.* (1980).

1. Test for significant differences between the five species in terms of the mean values and the variation in the nine variables. Test

Table 4.4 Values for nine mandible measurements for samples of five canine groups. The variables are X_1 = length of mandible, X_2 = breadth of mandible below 1st molar, X_3 = breadth of articular condyle, X_4 = height of mandible below 1st molar, X_5 = length of 1st molar, X_6 = breadth of first molar, X_7 = length of 1st to 3rd molar inclusive (1st to 2nd for cuon), X_8 = length from 1st to 4th premolar inclusive, and X_9 = breadth of lower canine, all measured in mm

	X_1	X_2	X_3	X_4	X_5	X_6	X_7	X_8	X_9	Sex
Modern dogs from Thailand										
1	123	10.1	23	23	19	7.8	32	33	5.6	M
2	127	9.6	19	22	19	7.8	32	40	5.8	M
3	121	10.2	18	21	21	7.9	35	38	6.2	M
4	130	10.7	24	22	20	7.9	32	37	5.9	M
5	149	12.0	25	25	21	8.4	35	43	6.6	M
6	125	9.5	23	20	20	7.8	33	37	6.3	M
7	126	9.1	20	22	19	7.5	32	35	5.5	M
8	125	9.7	19	19	19	7.5	32	37	6.2	M
9	121	9.6	22	20	18	7.6	31	35	5.3	F
10	122	8.9	20	20	19	7.6	31	35	5.7	F
11	115	9.3	19	19	20	7.8	33	34	6.5	F
12	112	9.1	19	20	19	6.6	30	33	5.1	F
13	124	9.3	21	21	18	7.1	30	36	5.5	F
14	128	9.6	22	21	19	7.5	32	38	5.8	F
15	130	8.4	23	20	19	7.3	31	40	5.8	F
16	127	10.5	25	23	20	8.7	32	35	6.1	F
Golden jackals										
1	120	8.2	18	17	18	7.0	32	35	5.2	M
2	107	7.9	17	17	20	7.0	32	34	5.3	M
3	110	8.1	18	16	19	7.1	31	32	4.7	M
4	116	8.5	20	18	18	7.1	32	33	4.7	M
5	114	8.2	19	18	19	7.9	32	33	5.1	M
6	111	8.5	19	16	18	7.1	30	33	5.0	M
7	113	8.5	17	18	19	7.1	30	34	4.6	M
8	117	8.7	20	17	18	7.0	30	34	5.2	M
9	114	9.4	21	19	19	7.5	31	35	5.3	M
10	112	8.2	19	17	19	6.8	30	34	5.1	M
11	110	8.5	18	17	19	7.0	31	33	4.9	F
12	111	7.7	20	18	18	6.7	30	32	4.5	F
13	107	7.2	17	16	17	6.0	28	35	4.7	F
14	108	8.2	18	16	17	6.5	29	33	4.8	F
15	110	7.3	19	15	17	6.1	30	33	4.5	F
16	105	8.3	19	17	17	6.5	29	32	4.5	F
17	107	8.4	18	17	18	6.2	29	31	4.3	F
18	106	7.8	19	18	18	6.2	31	32	4.4	F
19	111	8.4	17	16	18	7.0	30	34	4.7	F
20	111	7.6	19	17	18	6.5	30	35	4.6	F

Table 4.4 (*Contd.*)

	X_1	X_2	X_3	X_4	X_5	X_6	X_7	X_8	X_9	Sex
Cuons										
1	123	9.7	22	21	20	7.8	27	36	6.1	M
2	135	11.8	25	21	23	8.9	31	38	7.1	M
3	138	11.4	25	25	22	9.0	30	38	7.3	M
4	141	10.8	26	25	21	8.1	29	39	6.6	M
5	135	11.2	25	25	21	8.5	29	39	6.7	M
6	136	11.0	22	24	22	8.1	31	39	6.8	M
7	131	10.4	23	23	23	8.7	30	36	6.8	M
8	137	10.6	25	24	21	8.3	28	38	6.5	M
9	135	10.5	25	25	21	8.4	29	39	6.9	M
10	131	10.9	25	24	21	8.5	29	35	6.2	F
11	130	11.3	22	23	21	8.7	29	37	7.0	F
12	144	10.8	24	26	22	8.9	30	42	7.1	F
13	139	10.9	26	23	22	8.7	30	39	6.9	F
14	123	9.8	23	22	20	8.1	26	34	5.6	F
15	137	11.3	27	26	23	8.7	30	39	6.5	F
16	128	10.0	22	23	22	8.7	29	37	6.6	F
17	122	9.9	22	22	20	8.2	26	36	5.7	F
Indian wolves										
1	167	11.5	29	28	25	9.5	41	45	7.2	M
2	164	12.3	27	26	25	10.0	42	47	7.9	M
3	150	11.5	21	24	25	9.3	41	46	8.5	M
4	145	11.3	28	24	24	9.2	36	41	7.2	M
5	177	12.4	31	27	27	10.5	43	50	7.9	M
6	166	13.4	32	27	26	9.5	40	47	7.3	M
7	164	12.1	27	24	25	9.9	42	45	8.3	M
8	165	12.6	30	26	25	7.7	40	43	7.9	M
9	131	11.8	20	24	23	8.8	38	40	6.5	F
10	163	10.8	27	24	24	9.2	39	48	7.0	F
11	164	10.7	24	23	26	9.5	43	47	7.6	F
12	141	10.4	20	23	23	8.9	38	43	6.0	F
13	148	10.6	26	21	24	8.9	39	40	7.0	F
14	158	10.7	25	25	24	9.8	41	45	7.4	F
Prehistoric Thai dogs										
1	112	10.1	17	18	19	7.7	31	33	5.8	?
2	115	10.0	18	23	20	7.8	33	36	6.0	?
3	136	11.9	22	25	21	8.5	36	39	7.0	?
4	111	9.9	19	20	18	7.3	29	34	5.3	?
5	130	11.2	23	27	20	9.1	35	35	6.6	?
6	125	10.7	19	26	20	8.4	33	37	6.3	?
7	132	9.6	19	20	19	9.7	35	38	6.6	?
8	121	10.7	21	23	19	7.9	32	35	6.0	?
9	122	9.8	22	23	18	7.9	32	35	6.1	?
10	124	9.5	20	24	19	7.6	32	37	6.0	?

both for overall differences and for differences between the prehistoric Thai dogs and each of the other groups singly. What conclusion do you draw with regard to the similarity between prehistoric Thai dogs and the other groups?
2. Is there evidence of differences between the size of males and females of the same species for the first four groups?
3. Using a suitable graphical method, compare the distribution of the nine variables for the prehistoric and modern Thai dogs.

References

Carter, E.M., Khatri, C.G. and Srivastava, M.S. (1979) The effect of inequality of variances on the *t*-test. *Sankhya* **41**, 216–25.

Dixon, W.J. and Massey, F.J. (1969) *Introduction to Statistical Analysis*, 3rd edn. McGraw-Hill, New York.

Higham, C.F.W., Kijngam, A. and Manly, B.F.J. (1980) An analysis of prehistoric canid remains from Thailand. *Journal of Archaeological Science* **7**, 149–65.

Levene, H. (1960) Robust tests for equality of variance. In *Contributions to Probability and Statistics* (eds I. Olkin, S.G. Ghurye, W. Hoeffding, W.G. Madow and H.B. Mann), pp. 278–92. Stanford University Press, California.

Lotus Development Corporation (1992) LOTUS 123, Release 2.4. Lotus Development Corporation, 55 Cambridge Parkway, Cambridge, Massachusetts.

Schultz, B. (1983) On Levene's test and other statistics of variation. *Evolutionary Theory* **6**, 197–203.

Srivastava, M.S. and Carter, E.M. (1983) *An Introduction to Applied Multivariate Statistics*. North-Holland, New York.

Van Valen, L. (1978) The statistics of variation. *Evolutionary Theory* **4**, 33–43. (Erratum *Evolutionary Theory* **4**, 202.)

Yao, Y. (1965) An approximate degrees of freedom solution to the multivariate Behrens–Fisher problem. *Biometrika* **52**, 139–47.

CHAPTER FIVE

Measuring and testing multivariate distances

5.1 Multivariate distances

A large number of multivariate problems can be viewed in terms of 'distances' between single observations, or between samples of observations, or between populations of observations. For example, considering the data in Table 1.4 on mandible measurements of dogs, wolves, jackals, cuons and dingos, it is sensible to ask how far one of these groups is from the other six groups. The idea then is that if two animals have similar mean mandible measurements then they are 'close', whereas if they have rather different mean measurements then they are 'distant' from each other. Throughout this chapter it is this concept of 'distance' that is being used.

A large number of distance measures have been proposed and used in multivariate analyses. Here only some of the most common ones will be mentioned. It is fair to say that measuring distances is a topic where a certain amount of arbitrariness seems unavoidable.

A possible situation is that there are n objects being considered, with a number of measurements being taken on each of these, and the measurements are of two types. For example, in Table 1.3 results are given for four environmental variables and six gene frequencies for 16 colonies of a butterfly. Two sets of distances can therefore be calculated between the colonies. One set can be environmental distances and the other set genetic distances. An interesting question is then whether there is a significant relationship between these two sets of distances. Mantel's (1967) test which is described in section 5.5 is useful in this context.

5.2 Distances between individual observations

To begin the discussion on measuring distances, consider the simplest case where there are n objects, each of which has values for p

variables, $X_1, X_2, \ldots, X_p$. The values for object i can then be denoted by $x_{i1}, x_{i2}, \ldots, x_{ip}$ and those for object j by $x_{j1}, x_{j2}, \ldots, x_{jp}$. The problem is to measure the 'distance' between these two objects.

If there are only $p = 2$ variables then the values can be plotted as shown in Fig. 5.1. Pythagoras' theorem then says that the length, d_{ij}, of the line joining the point for object i to the point for object j (the *Euclidean distance*) is

$$d_{ij} = \sqrt{\{(x_{i1} - x_{j1})^2 + (x_{i2} - x_{j2})^2\}}.$$

With $p = 3$ variables the values can be taken as the coordinates in space for plotting the positions of indviduals i and j (Fig. 5.2). Pythagoras' theorem then gives the distance between the two points

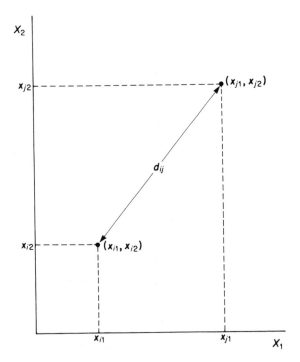

Figure 5.1 The Euclidean distance between objects i and j, with $p = 2$ variables.

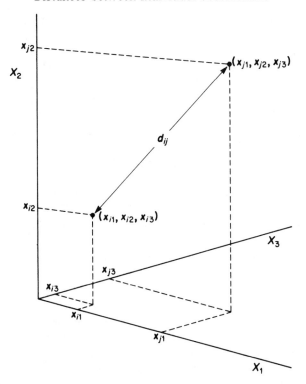

Figure 5.2 The Euclidean distance between objects i and j, with $p = 3$ variables.

as

$$d_{ij} = \sqrt{\left\{ \sum_{k=1}^{3} (x_{ik} - x_{jk})^2 \right\}}.$$

With more than three variables it is not possible to use variable values as the coordinates for physically plotting points. However, the two- and three-variable cases suggest that the generalized Euclidean distance

$$d_{ij} = \sqrt{\left\{ \sum_{k=1}^{p} (x_{ik} - x_{jk})^2 \right\}} \tag{5.1}$$

may serve as a satisfactory measure for many purposes.

From the form of equation (5.1) it is clear that if one of the variables measured is much more variable than the others then this will dominate the calculation of distances. For example, to take an extreme case, suppose that n men are being compared and that X_1 is their stature and the other variables are tooth dimensions, with all the measurements being in millimetres. Stature differences will then be in the order of perhaps 20 or 30 mm while tooth dimension differences will be in the order of 1 or 2 mm. Simple calculations of d_{ij} will then provide distances between individuals that are essentially stature differences only, with tooth differences having negligible effects. Clearly there will be a scaling problem.

In practice it is usually desirable for all variables to have about the same influence on the distance calculation. This is achieved by a preliminary scaling of the variables to standardize them. This can be done, for example, by dividing each variable by its standard deviation for the n individuals being compared.

Example 5.1 Distances between dogs and related species

As an example of the use of the Euclidean distance measure, consider the data in Table 1.4 for mean mandible measurements of seven groups of dogs and related species. It may be recalled from Chapter 1 that the main question with these data is how the prehistoric dogs relate to the other groups.

The first step in calculating distances is to standardize the measure-meants. Here this will be done by expressing them as deviations from means in units of standard deviations. For example, the first

Table 5.1 Standardized variable values calculated from the original data in Table 1.4

X_1	X_2	X_3	X_4	X_5	X_6
−0.50	−0.50	−0.74	−0.74	−0.49	−0.61
−1.52	−1.93	−1.12	−1.39	−0.86	−1.31
1.92	1.60	1.84	1.95	1.68	1.62
0.65	0.60	1.04	0.74	1.26	0.95
0.14	0.33	−0.04	0.00	−1.19	−0.40
−0.56	0.03	−0.14	−0.19	0.03	0.66
−0.12	−0.13	−0.84	−0.37	−0.43	−0.90

Table 5.2 Euclidean distances between seven animal groups

	Modern dog	Golden jackal	Chinese wolf	Indian wolf	Cuon	Dingo	Prehistoric dog
Modern dog	–						
Golden jackal	2.07	–					
Chinese wolf	5.81	7.69	–				
Indian wolf	3.66	5.49	2.31	–			
Cuon	1.63	3.45	4.94	3.14	–		
Dingo	1.68	3.44	4.55	2.37	1.80	–	
Prehistoric dog	0.72	2.58	5.52	3.49	1.38	1.84	–

measurement X_1 (breadth) has a mean of 10.486 mm and a standard deviation of 1.572 mm for the seven groups. The standardized variable values are then calculated as follows: modern dog, $(9.7 - 10.486)/1.572 = -0.50$; golden jackal, $(8.1 - 10.486)/1.572 = -1.52;\ldots;$ prehistoric dog, $(10.3 - 10.486)/1.572 = -0.12$. Standardized values for all the variables are shown in Table 5.1.

Using equation (5.1) the distances shown in Table 5.2 have been calculated from the standardized variables. It is clear that the prehistoric dogs are rather similar to modern dogs in Thailand. Indeed, the distance between these two groups is the smallest distance in the whole table. (Higham *et al.* (1980) concluded from a more complicated analysis that the modern and prehistoric dogs are indistinguishable.)

5.3 Distances between populations and samples

A number of measures have been proposed for the distance between two multivariate populations when information is available on the means, variances and covariances of the populations. Here only two will be considered.

Suppose that g populations are available and the multivariate distributions in these populations are known for p variables X_1, $X_2, \ldots, X_p$. Let the mean of variable X_k in the ith population be μ_{ki}, and assume that the variance of X_k is the same value, V_k, in all the populations. Penrose (1953) proposed the relatively simple measure

$$P_{ij} = \sum_{k=1}^{p} \frac{(\mu_{ki} - \mu_{kj})^2}{pV_k} \tag{5.2}$$

for the distance between population i and population j.

A disadvantage of Penrose's measure is that it does not take into account the correlations between the p variables. This means that when two variables are measuring essentially the same thing, and hence are highly correlated, they still individually both contribute about the same amount to population distances as a third variable that is independent of all other variables.

A measure that does take into account correlations between variables is the Mahalanobis (1948) distance,

$$D_{ij}^2 = \sum_{r=1}^{p} \sum_{s=1}^{p} (\mu_{ri} - \mu_{rj})v^{rs}(\mu_{si} - \mu_{sj}), \tag{5.3}$$

where v^{rs} is the element in the rth row and sth column of the inverse of the covariance matrix for the p variables. This is a quadratic form that can be written in the alternative way

$$D_{ij}^2 = (\pmb{\mu}_i - \pmb{\mu}_j)' \mathbf{V}^{-1} (\pmb{\mu}_i - \pmb{\mu}_j), \qquad (5.4)$$

where

$$\pmb{\mu}_i = \begin{bmatrix} \mu_{1i} \\ \mu_{2i} \\ \vdots \\ \mu_{pi} \end{bmatrix}$$

is the vector of means for the ith population and $\mathbf{V}$ is the covariance matrix. This measure can only be calculated if the population covariance matrix is the same for all populations.

The Mahalanobis distance is frequently used to measure the distance of a single multivariate observation from the centre of the population that the observation comes from. If $x_1, x_2, \ldots, x_p$ are the values of $X_1, X_2, \ldots, X_p$ for the individual, with corresponding population mean values of $\mu_1, \mu_2, \ldots, \mu_p$, then

$$D^2 = \sum_{r=1}^{p} \sum_{s=1}^{p} (x_r - \mu_r) v^{rs} (x_s - \mu_s)$$

$$= (\mathbf{x} - \pmb{\mu})' \mathbf{V}^{-1} (\mathbf{x} - \pmb{\mu}), \qquad (5.5)$$

where $\mathbf{x}' = (x_1, x_2, \ldots, x_p)$ and $\pmb{\mu}' = (\mu_1, \mu_2, \ldots, \mu_p)$. As before $\mathbf{V}$ denotes the population covariance matrix and v^{rs} is the element in the rth row and sth column of $\mathbf{V}^{-1}$.

The value of D^2 can be thought of as a multivariate residual for the observation $\mathbf{x}$. A 'residual' here means a measure of how far the observation $\mathbf{x}$ is from the centre of the distributions of all values, taking into account all the variables being considered. An important and useful result is that if the population being considered is multivariate normally distributed, then the values of D^2 will follow a chi-squared distribution with p degrees of freedom. A significantly large value of D^2 means that the corresponding observation is either (a) a genuine but unlikely record, or (b) a record containing some mistake. This suggests that a check should be made to see that the observation is correct and does not include an error.

The equations (5.2) to (5.5) can obviously be used with sample data if estimates of population means, variances and covariances are used in place of true values. In that case the covariance matrix V involved in equations (5.3) and (5.4) should be replaced with the pooled estimate from all the samples available. To be precise, suppose that there are m samples, with the ith sample being of size n_i, with a sample covariance matrix of C_i. Then it is appropriate to take

$$C = \sum_{i=1}^{m} (n_i - 1)C_i \Big/ \sum_{i=1}^{m} (n_i - 1) \tag{5.6}$$

as the pooled estimate of the common covariance matrix. The single-sample covariance matrix C_i is said to have $n_i - 1$ degrees of freedom, while C has a total of $\sum(n_i - 1)$ degrees of freedom. The sample covariance matrices should be calculated using equations (2.2), (2.3) and (2.7).

In principle the Mahalanobis distance is superior to the Penrose distance because it uses information on covariances. However, this advantage is only present when covariances are accurately known. When covariances can only be estimated with a small number of degrees of freedom it is probably best to use the simpler Penrose measure. It is difficult to say precisely what a 'small number of degrees of freedom' means in this context. Certainly there should be no problem with using Mahalanobis distances based on a covariance matrix with the order of 100 or more degrees of freedom.

Example 5.2 Distances between samples of Egyptian skulls

For the five samples of male Egyptian skulls shown in Table 1.2 the mean vectors are as follows:

$$\bar{x}_1 = \begin{bmatrix} 131.37 \\ 133.60 \\ 99.17 \\ 50.53 \end{bmatrix}, \quad \bar{x}_2 = \begin{bmatrix} 132.37 \\ 132.70 \\ 99.07 \\ 50.23 \end{bmatrix}, \quad \bar{x}_3 = \begin{bmatrix} 134.47 \\ 133.80 \\ 96.03 \\ 50.57 \end{bmatrix},$$

$$\bar{x}_4 = \begin{bmatrix} 135.50 \\ 132.30 \\ 94.53 \\ 51.97 \end{bmatrix}, \quad \text{and} \quad \bar{x}_5 = \begin{bmatrix} 136.17 \\ 130.33 \\ 93.50 \\ 51.37 \end{bmatrix},$$

while the covariance matrices, calculated as indicated by equation (2.7), are

$$
\mathbf{C}_1 = \begin{bmatrix} 26.31 & 4.15 & 0.45 & 7.25 \\ 4.15 & 19.97 & -0.79 & 0.39 \\ 0.45 & -0.79 & 34.63 & -1.92 \\ 7.25 & 0.39 & -1.92 & 7.64 \end{bmatrix},
$$

$$
\mathbf{C}_2 = \begin{bmatrix} 23.14 & 1.01 & 4.77 & 1.84 \\ 1.01 & 21.60 & 3.37 & 5.62 \\ 4.77 & 3.37 & 18.89 & 0.19 \\ 1.84 & 5.62 & 0.19 & 8.74 \end{bmatrix},
$$

$$
\mathbf{C}_3 = \begin{bmatrix} 12.12 & 0.79 & -0.78 & 0.90 \\ 0.79 & 24.79 & 3.59 & -0.09 \\ -0.78 & 3.59 & 20.72 & 1.67 \\ 0.90 & -0.09 & 1.67 & 12.60 \end{bmatrix},
$$

$$
\mathbf{C}_4 = \begin{bmatrix} 15.36 & -5.53 & -2.17 & 2.05 \\ -5.53 & 26.36 & 8.11 & 6.15 \\ -2.17 & 8.11 & 21.09 & 5.33 \\ 2.05 & 6.15 & 5.33 & 7.96 \end{bmatrix},
$$

$$
\text{and } \mathbf{C}_5 = \begin{bmatrix} 28.63 & -0.23 & -1.88 & -1.99 \\ -0.23 & 24.71 & 11.72 & 2.15 \\ -1.88 & 11.72 & 25.57 & 0.40 \\ -1.99 & 2.15 & 0.40 & 13.83 \end{bmatrix}.
$$

Although the five sample covariance matrices appear to differ somewhat, it has been shown in Example 4.3 that the differences are not significant. It is therefore reasonable to pool them using equation (5.6). Because the sample sizes are all 30 this just amounts to taking the average of the five matrices, which is

$$
\mathbf{C} = \begin{bmatrix} 21.111 & 0.037 & 0.079 & 2.009 \\ 0.037 & 23.485 & 5.200 & 2.845 \\ 0.079 & 5.200 & 24.179 & 1.133 \\ 2.009 & 2.845 & 1.133 & 10.153 \end{bmatrix}
$$

with $\sum(n_i - 1) = 145$ degrees of freedom.

Penrose's distance measures of equation (5.2) can now be calculated between each pair of samples. There are $p = 4$ variables with variances that are estimated by $\hat{V}_1 = 21.111$, $\hat{V}_2 = 23.485$, $\hat{V}_3 = 24.179$ and $\hat{V}_4 = 10.153$, these being the diagonal terms in the pooled covariance matrix. The sample mean values given in the vectors $\bar{x}_1$ to $\bar{x}_5$ are estimates of population means. For example, the distance between sample 1 and sample 2 is calculated as

$$P_{12} = \frac{(131.37 - 132.37)^2}{4 \times 21.111} \times \frac{(133.60 - 132.70)^2}{4 \times 23.485}$$

$$+ \frac{(99.17 - 99.07)^2}{4 \times 24.179} + \frac{(50.53 - 50.23)^2}{4 \times 10.153}$$

$$= 0.023.$$

This only has meaning in comparison with the distances between the other pairs of samples. Calculating these as well provides the following distance matrix:

	Early predynastic	Late predynastic	12/13th dynasties	Ptolemaic	Roman
Early predynastic	–				
Late predynastic	0.023	–			
12/13th dynasties	0.216	0.163	–		
Ptolemaic	0.493	0.404	0.108	–	
Roman	0.736	0.583	0.244	0.066	–

It will be recalled from Example 4.3 that the mean values change significantly from sample to sample. The Penrose distances show that the changes are cumulative over time: the samples that are closest in time are relatively similar whereas the samples that are far apart in time are very different.

Turning next to Mahalanobis distances, these can be calculated from equation (5.3). The inverse of the pooled covariance matrix $\mathbf{C}$ is

$$\mathbf{C}^{-1} = \begin{bmatrix} 0.0483 & 0.0011 & 0.0001 & -0.0099 \\ 0.0011 & 0.0461 & -0.0094 & -0.0121 \\ 0.0001 & -0.0094 & 0.0435 & -0.0022 \\ -0.0099 & -0.0121 & -0.0022 & 0.1041 \end{bmatrix}.$$

Using this and the sample means gives the distance from sample 1 to sample 2 to be

$$D_{12}^2 = (131.37 - 132.37)0.0483(131.37 - 132.37)$$
$$+ (131.37 - 132.37)0.0011(133.60 - 132.70)$$
$$+ \cdots - (50.53 - 50.23)0.0022(99.17 - 99.07)$$
$$+ (50.53 - 50.23)0.1041(50.53 - 50.23)$$
$$= 0.091.$$

Calculating the other distances between samples in the same way provides the distance matrix:

	Early pre-dynastic	Late pre-dynastic	12/13th dynasties	Ptolemaic	Roman
Early predynastic	–				
Late predynastic	0.091	–			
12/13th dynasties	0.903	0.729	–		
Ptolemaic	1.881	1.594	0.443	–	
Roman	2.697	2.176	0.911	0.219	–

A comparison between these distances and the Penrose distances shows a very good agreement. The Mahalanobis distances are three to four times as great as the Penrose distances. However, the relative distances between samples are almost the same for both measures. For example, the Penrose measure suggest that the distance from the early predynastic sample to the Roman sample is $0.736/0.023 = 32.0$ times as great as the distance from the early predynastic to the late predynastic sample. The corresponding ratio for the Mahalanobis measure is $2.697/0.091 = 29.6$.

5.4 Distances based upon proportions

A particular situation that sometimes occurs is that the variables being used to measure the distance between populations or samples are proportions whose sum is unity. For example, the animals of a certain species might be classified into K genetic classes. One colony might then have proportions p_1 of class 1, p_2 of class 2,..., p_K of class K, while a second colony has proportions q_1 of class 1, q_2 of class $2, \ldots, q_k$ of class K. The question then arises of how different the colonies are in genetic terms.

Various indices of distance have been proposed with this type of proportion data. For example,

$$d_1 = \sum_{i=1}^{K} |p_i - q_i|/2, \qquad (5.7)$$

which is half of the sum of absolute proportion differences, is one possibility. This takes the value 1 when there is no overlap of classes and the value 0 when $p_i = q_i$ for all i. Another possibility is

$$d_2 = 1 - \sum_{i=1}^{K} p_1 q_i \bigg/ \sqrt{\left\{ \sum_{i=1}^{K} p_i^2 \sum_{i=1}^{K} q_i^2 \right\}}, \qquad (5.8)$$

which again varies from 1 (no overlap) to 0 (equal proportions).

Because d_1 and d_2 vary from 0 to 1, it follows that $1 - d_1$ and $1 - d_2$ are measures of the similarity between the items being compared. In fact, it is in terms of similarities that the indices are often used. For example,

$$s_1 = 1 - d_1 = 1 - \sum_{i=1}^{K} |p_i - q_i|/2$$

is often used as a measure of the niche overlap between two species, where p_i is the fraction of the resources used by species 1 that are of type i and q_i is the fraction of the resources used by species 2 that are of type i. Then $s_1 = 0$ indicates that the two species use completely different resources, and $s_1 = 1$ indicates that the two species use exactly the same resources.

A similarity measure can also be constructed from any distance measure D that varies from zero to infinity. Taking $S = 1/D$ gives a similarity that ranges from infinity for two items that are no distance apart, to 0 for two objects that are infinitely distant. Alternatively, $1/(1 + D)$ ranges from 1 when $D = 0$, to 0 when D is infinite.

5.5 Presence–absence data

Another common situation is where the similarity or distance between two items must be based on a list of their presences and absences. For example, their might be interest in the similarity between two plant species in terms of their distributions at 10 sites. The data might then take the following form, where 1 indicates presence and 0 indicates absence:

Site	1	2	3	4	5	6	7	8	9	10
Species 1	0	0	1	1	1	0	1	1	1	0
Species 2	1	1	1	1	0	0	0	0	1	1

Such data are often summarized in the form shown in Table 5.3 as counts of the number of times that both species are present (a), only one species is present (b and c), or both species are absent (d). Thus for the present case $a = 3$, $b = 3$, $c = 3$ and $d = 1$.

In this situation some of the commonly used similarity measures are the simple matching index $(a+d)/n$, the Ochiai index $a/\sqrt{\{(a+b)(a+c)\}}$, the Dice index $2a/(2a+b+c)$ and the Jaccard index $a/(a+b+c)$. These all vary from 0 (no similarity) to 1 (complete similarity), so that complementary distance measures can be calculated by substracting the similarity indices from 1. There has been some debate about whether the number of joint absences (d) should be used in the calculation because of the danger of concluding that two species are similar simply because they are both absent from many sites. This is certainly a valid point in many situations, and suggests that the simple matching index should be used with caution.

Table 5.3 Presence and absence data obtained for two species at n sites

	Species 2		
	Present	*Absent*	*Total*
Species 1			
Present	a	b	$a+b$
Absent	c	d	$c+d$
	$a+c$	$b+d$	n

5.6   The Mantel test

A useful test for comparing two distance or similarity matrices was introduced by Mantel (1967) as a solution to the problem of detecting space and time clustering of diseases.

To understand the nature of the procedure the following simple example is helpful. Suppose that four objects are being studied, and that two sets of variables have been measured for each of these. The first set of variables can then be used to construct a 4×4 matrix where the entry in the ith row and jth column is the 'distance' between object i and object j. The distance matrix might be, for example,

$$\mathbf{M} = \begin{bmatrix} m_{11} & m_{12} & m_{13} & m_{14} \\ m_{21} & m_{22} & m_{23} & m_{24} \\ m_{31} & m_{32} & m_{33} & m_{34} \\ m_{41} & m_{42} & m_{43} & m_{44} \end{bmatrix} = \begin{bmatrix} 0.0 & 1.0 & 1.4 & 0.9 \\ 1.0 & 0.0 & 1.1 & 1.6 \\ 1.4 & 1.1 & 0.0 & 0.7 \\ 0.9 & 1.6 & 0.7 & 0.0 \end{bmatrix}.$$

It is symmetric because, for example, the distance from object 2 to object 3 must be the same as the distance from object 3 to object 2 (1.1 units). Diagonal elements are zero because these represent distances from objects to themselves.

The second set of variables can also be used to construct a matrix of distances between the objects. For the example this will be taken as

$$\mathbf{E} = \begin{bmatrix} e_{11} & e_{12} & e_{13} & e_{14} \\ e_{21} & e_{22} & e_{23} & e_{24} \\ e_{31} & e_{32} & e_{33} & e_{34} \\ e_{41} & e_{42} & e_{43} & e_{44} \end{bmatrix} = \begin{bmatrix} 0.0 & 0.5 & 0.8 & 0.6 \\ 0.5 & 0.0 & 0.5 & 0.9 \\ 0.8 & 0.5 & 0.0 & 0.4 \\ 0.6 & 0.9 & 0.4 & 0.0 \end{bmatrix}.$$

Like $\mathbf{M}$, this is symmetric with zeros down the diagonal.

Mantel's test is concerned with assessing whether the elements in $\mathbf{M}$ and $\mathbf{E}$ show correlation. Assuming $n \times n$ matrices, the test statistic

$$Z = \sum_{i=2}^{n} \sum_{j=2}^{i-1} m_{ij}e_{ij} \tag{5.9}$$

is calculated and compared with the distribution of Z that is obtained by taking the objects in a random order for one of the matrices.

That is to say, matrix **M** can be left as it is. A random order can then be chosen for the objects for matrix **E**. For example, suppose that a random ordering of objects turns out to be 3,2,4,1. This then gives a randomized **E** matrix of

$$\mathbf{E_R} = \begin{bmatrix} 0.0 & 0.5 & 0.4 & 0.8 \\ 0.5 & 0.0 & 0.9 & 0.5 \\ 0.4 & 0.9 & 0.0 & 0.6 \\ 0.8 & 0.5 & 0.6 & 0.0 \end{bmatrix}.$$

The entry in row 1, column 2 is 0.5, the distance between objects 3 and 2; the entry in row 1, column 3 is 0.4, the distance between objects 3 and 4; and so on. A Z value can be calculated using **M** and $\mathbf{E_R}$. Repeating this procedure using different random orders of the objects for $\mathbf{E_R}$ produces the randomized distribution of Z. A check can then be made to see whether the observed Z value is a typical value from this distribution.

The basic idea is that if the two measures of distance are quite unrelated then the matrix **E** will be just like one of the randomly ordered matrices $\mathbf{E_R}$. Hence the observed Z will be a typical randomized Z value. On the other hand, if the two distance measures have a positive correlation then the observed Z will tend to be larger than values given by randomization. A negative correlation between distances should not occur but if it does then the result will be that the observed Z value will tend to be low when compared to the randomized distribution.

With n objects there are $n!$ different possible orderings of the object numbers. There are therefore $n!$ possible randomizations of the elements of **E**, some of which might give the same Z values. Hence in our example with four objects the randomized Z distribution has $4! = 24$ equally likely values. It is not too difficult to calculate all of these. More realistic cases might involve, say, 15 objects, in which case the number of possible Z values is $15! \simeq 1.3 \times 10^{12}$. Enumerating all of these then becomes impractical and there are two possible approaches for carrying out the Mantel test. A large number of randomized $\mathbf{E_R}$ matrices can be generated on the computer and the resulting distribution of Z values used in place of the true randomized distribution. Alternatively, the mean, $E(Z)$, and variance var(Z), of the randomized distribution of Z can be calculated, and

$$g = [Z - E(Z)]/[\text{var}(Z)]^{1/2}$$

can be treated as a standard normal variate.

Mantel (1967) provided formulae for the mean and variance of Z in the null hypothesis case of no correlation between the distance measures. There is, however, some doubt about the validity of the normal approximation for the test statistic g (Mielke, 1978). Given the ready availability of computers it therefore seems best to perform randomizations rather than to rely on this approximation.

The test statistic Z of equation (5.9) is the sum of the products of the elements in the lower diagonal parts of the matrices $\mathbf{M}$ and $\mathbf{E}$. The only reason for using this particular statistic is that Mantel's equations for the mean and variance are available. However, if it is decided to determine significance by computer and randomizations there is no particular reason why the test statistic should not be changed. Indeed, values of Z are not particularly informative except in comparison with the mean and variance. It may therefore be more useful to take the correlation between the lower diagonal elements of $\mathbf{M}$ and $\mathbf{E}$ as the test statistic instead of Z. This correlation is

$$r = \frac{Z - n(n-1)\bar{m}\bar{e}/2}{\sqrt{\left\{ \left(\sum_{i=2}^{n} \sum_{j=1}^{i-1} m_{ij}^2 - n(n-1)\bar{m}^2/2 \right) \left(\sum_{i=2}^{n} \sum_{j=1}^{i-1} e_{ij}^2 - n(n-1)\bar{e}^2/2 \right) \right\}}} \tag{5.10}$$

where $\bar{m}$ is the mean of the m_{ij} values, $\bar{e}$ is the mean of the e_{ij} values, and $n(n-1)/2$ is the number of lower diagonal elements in the matrices. This statistic has the usual interpretation in terms of the relationship between the two distance measures. Thus r lies in the range -1 to $+1$, with $r = -1$ indicating a perfect negative correlation, $r = 0$ indicating no correlation, and $r = +1$ indicating a perfect positive correlation. The significance or otherwise of the data will be the same for the test statistics Z and r because r is just Z with a constant subtracted, divided by another constant.

Example 5.3 More on distances between samples of Egyptian skulls

Returning to the Egyptian skull data, we can ask the question of whether the distances given in Example 5.2, based upon four skull measurements, are significantly related to the time differences between the five samples. This certainly does seem to be the case but a definitive answer is provided by Mantel's test.

The sample times are approximately 4000 BC (early predynastic), 3300 BC (late predynastic), 1850 BC (12th and 13th dynasties), 200 BC (Ptolemaic), and AD 150 (Roman). Comparing Penrose's distance measures with time differences (in thousands of years) therefore provides the following lower diagonal distance matrices between the samples:

Penrose's distances					*Time distances*				
–					–				
0.023	–				0.70	–			
0.216	0.163	–			2.15	1.45			
0.493	0.404	0.108	–		3.80	3.10	1.65	–	
0.736	0.583	0.244	0.066	–	4.15	3.45	2.00	0.35	–

The correlation between the elements of these matrices is 0.954. It appears, therefore, that the distances agree very well.

There are $5! = 120$ possible ways to reorder the five samples for one of the two matrices and, consequently, there are 120 elements in the randomization distribution for the correlation. Of these, one is the observed correlation of 0.954 and another is a larger correlation. It follows that the observed correlation is significantly high at the $(2/120)100\% = 1.7\%$ level. There is clear evidence of a relationship between the two distance matrices. A one-sided test is appropriate because there is no reason why the skulls should become more similar as they get further apart in time.

The matrix correlation between Mahalanobis distances and time distance is 0.964. This is also significantly large at the 1.7% level when compared with the full randomization distribution.

5.7 Computational methods and computer programs

The calculation of distance and similarity measures is the first step in the analysis of multivariate data using cluster analysis and ordination methods. For this reason the calculation of these measures is often easiest to do using computer programs that are designed for these methods, or the clustering and ordination options of more general statistical packages. For example, the program MVSP (Kovach, 1993) allows 18 different measures to be computed from initial data matrices and saved for later use.

Computer programs for Mantel's test on distance and similarity matrices are not readily available. A FORTRAN subroutine for this purpose is provided by Manly (1991), and there is an option in the program RT (Manly, 1992) that can be used to carry out the test with the regression coefficient for the values of one matrix regressed against the values in the second matrix as the test statistic. This is exactly equivalent to using the Z value of equation (5.9) or the matrix correlation of equation (5.10) as the test statistic, but has the advantage of allowing a test for a relationship between the values in one dependent matrix and several other matrices, as discussed in the next section. The program RT also allows the construction of distance matrices using the measures of equations (5.1) and (5.7).

5.8 Further reading

The use of different measures of distance and similarity is the subject of continuing debates, indicating a lack of agreement about what is the best under different circumstances. The problem is that no measure is perfect and the conclusions from an analysis may depend to some extent on which of several reasonable measures is used. The situation depends very much on what the purpose is for calculating the distances or similarities, and the nature of the data available. For further information in relationship to cluster analysis and ordination see Romesburg (1984), Digby and Kempton (1987) and Ludwig and Reynolds (1988).

The usefulness of the Mantel randomization method for testing for an association between two distance or similarity matrices has led to a number of proposals for methods to analyse relationships between three or more such matrices. These are reviewed by Manly (1994). At present a major unresolved problem in this area relates to the question of how to take proper account of the effects of spatial correlation when, as is often the case, the items that distances and similarities are measured between tend to be similar when they are relatively close in space.

Exercise

Consider the data in Table 1.3.

1. Standardize the environmental variables altitude, annual precipitation, annual maximum temperature and annual minimum

temperature to means of zero and standard deviations of one, and calculate Euclidean distances between all pairs of colonies using equation (5.1) in order to obtain an environmental distance matrix.

2. Use the Pgi gene frequencies, converted to proportions, to calculate genetic distances between the colonies using equation (5.7).

3. Carry out a Mantel matrix randomization test to determine whether there is a significant positive relationship between the environmental and genetic distances and report your conclusions.

4. Explain why a significant positive relationship on a randomization test in a situation such as this could be the result of spatial correlations between the data for close colonies rather than from environmental effects on the genetic composition of colonies.

References

Digby, P.G.N. and Kempton, R.A. (1987) *Multivariate Analysis of Ecological Communities*. Chapman and Hall, London.

Higham, C.F.W., Kijngam, A. and Manly, B.F.J. (1980) Analysis of prehistoric canid remains from Thailand. *Journal of Archaeological Science* **7**, 149–65.

Kovach, W.L. (1993) *MVSP Plus, Version 2.1*. Kovach Computing Services, 85 Nant-y-Felin, Pentraeth, Anglesey LL75 8UY, Wales.

Ludwig, J.A. and Reynolds, J.F. (1988) *Statistical Ecology*. Wiley, New York.

Mahalanobis, P.C. (1948) Historic note on the D^2-statistic. *Sankhya* **9**, 237.

Manly, B.F.J. (1991) *Randomization and Monte Carlo Methods in Biology*. Chapman and Hall, London.

Manly, B.F.J. (1992) *RT, a Program for Randomization Testing*. Centre for Applications of Statistics and Mathematics, University of Otago, PO Box 56, Dunedin, New Zealand.

Manly, B.F.J. (1994) A review of computer intensive multivariate methods in ecology. In *Multivariate Analysis: Future Directions* (ed. C.R. Rao), pp. 307–46, Elsevier, Amsterdam.

Mantel, N. (1967) The detection of disease clustering and a generalized regression approach. *Cancer Research* **27**, 209–20.

Mielke, P.W. (1978) Classification and appropriate inferences for Mantel and Varland's nonparametric multivariate analysis technique. *Biometrics* **34**, 272–82.

Penrose, L.W. (1953) Distance, size and shape. *Annals of Eugenics* **18**, 337–43.

Romesburg, H.C. (1984) *Cluster Analysis for Researchers*. Lifetime Learning Publications, Belmont, California.

CHAPTER SIX

Principal components analysis

6.1 Definition of principal components

The technique of principal components analysis was first described by Karl Pearson (1901). He apparently believed that this was the correct solution to some of the problems that were of interest to biometricians at that time, although he did not propose a practical method of calculation for more than two or three variables. A description of practical computing methods came much later from Hotelling (1933). Even then the calculations were extremely daunting for more than a few variables because they had to be done by hand. It was not until electronic computers became widely available that the technique achieved widespread use.

Principal components analysis is one of the simplest of the multivariate methods that will be described in this book. The object of the analysis is to take p variables $X_1, X_2, \ldots, X_p$ and find combinations of these to produce indices $Z_1, Z_2, \ldots, Z_p$ that are uncorrelated. The lack of correlation is a useful property because it means that the indices are measuring different 'dimensions' in the data. However, the indices are also ordered so that Z_1 displays the largest amount of variation, Z_2 displays the second largest amount of variation, and so on. That is, $\text{var}(Z_1) \geqslant \text{var}(Z_2) \geqslant \cdots \geqslant \text{var}(Z_p)$, where $\text{var}(Z_i)$ denotes the variance of Z_i in the data set being considered. The Z_i are called the principal components. When doing a principal components analysis there is always the hope that the variances of most of the indices will be so low as to be negligible. In that case the variation in the data set can be adequately described by the few Z variables with variances that are not negligible. Some degree of economy is then achieved, and the variation in the p original X variables is accounted for by a smaller number of Z variables.

It must be stressed that a principal components analysis does not always work in the sense that a large number of original variables are reduced to a small number of transformed variables. Indeed, if the original variables are uncorrelated then the analysis does absolutely nothing. The best results are obtained when the original variables are very highly correlated, positively or negatively. If that is the case then it is quite conceivable that 20 or 30 original variables can be adequately represented by two or three principal components. If this desirable state of affairs does occur then the important principal components will be of some interest as measures of underlying 'dimensions' in the data. However, it will also be of value to know that there is a good deal of redundancy in the original variables, with most of them measuring similar things.

Before launching into a description of the calculations involved in a principal components analysis it may be of some value to look briefly at the outcome of the analysis when it is applied to the data in Table 1.1 on five body measurements of 49 female sparrows. Details of the analysis are given in Example 6.1. In this case the five measurements are quite highly correlated, as shown in Table 6.1. This is therefore good material for the analysis in question. In turns out, as we shall see, that the first principal component has a variance of 3.62 whereas the other components all have variances very much less than this (0.53, 0.39, 0.30 and 0.16). This means that the first principal component is by far the most important of the five components for representing the variation in the measurements of the 49 birds. The first component is calculated to be

$$Z_1 = 0.45X_1 + 0.46X_2 + 0.45X_3 + 0.47X_4 + 0.40X_5,$$

where $X_1, X_2, \ldots, X_5$ represent here the measurements in Table 1.1

Table 6.1 Correlations between the five body measurements of female sparrows calculated from the data of Table 1.1

Variable	X_1	X_2	X_3	X_4	X_5
X_1, total length	1.000				
X_2, alar extent	0.735	1.000			
X_3, length of beak & head	0.662	0.674	1.000		
X_4, length of humerus	0.645	0.769	0.763	1.000	
X_5, length of keel of sternum	0.605	0.529	0.526	0.607	1.000

after they have been standardized to have zero means and unit standard deviations. Clearly Z_1 is essentially just an average of the standardized body measurements and it can be thought of as a simple index of size. The analysis given in Example 6.1 leads to the conclusion that most of the differences between the 49 birds are a matter of size (rather than shape).

6.2 Procedure for a principal components analysis

A principal components analysis starts with data on p variables for n individuals, as indicated in Table 6.2. The first principal component is then the linear combination of the variables $X_1, X_2, \ldots, X_p$,

$$Z_1 = a_{11}X_1 + a_{12}X_2 + \cdots a_{1p}X_p$$

that varies as much as possible for the individuals, subject to the condition that

$$a_{11}^2 + a_{12}^2 + \cdots + a_{1p}^2 = 1.$$

Thus the variance of Z_1, var(Z_1), is as large as possible given this constraint on the constants a_{1j}. The constraint is introduced because if this is not done then var(Z_1) can be increased by simply increasing any one of the a_{1j} values. The second principal component,

$$Z_2 = a_{21}X_1 + a_{22}X_2 + \cdots + a_{2p}X_p,$$

is such that var(Z_2) is as large as possible subject to the constraint that

$$a_{21}^2 + a_{22}^2 + \cdots + a_{2p}^2 = 1,$$

Table 6.2 The form of data for a principal components analysis

Individual	X_1	X_2	$\cdots$	X_p
1	x_{11}	x_{12}	$\cdots$	x_{1p}
2	x_{21}	x_{22}	$\cdots$	x_{2p}
$\vdots$	$\vdots$	$\vdots$		$\vdots$
n	x_{n1}	x_{n2}	$\cdots$	x_{np}

and also to the condition that Z_1 and Z_2 are uncorrelated. The third principal component,

$$Z_3 = a_{31}X_1 + a_{32}X_2 + \cdots + a_{3p}X_p,$$

is such that var(Z_3) is as large as possible subject to the constraint that

$$a_{31}^2 + a_{32}^2 + \cdots + a_{3p}^2 = 1,$$

and also that Z_3 is uncorrelated with Z_2 and Z_1. Further principal components are defined by continuing in the same way. If there are p variables then there can be up to p principal components.

In order to use the results of a principal components analysis it is not necessary to know how the equations for the principal components are derived. However, it is useful to understand the nature of the equations themselves. In fact a principal components analysis just involves finding the eigenvalues of the sample covariance matrix.

The calculation of the sample covariance matrix has been described in Chapter 2. The important equations are (2.2), (2.3) and (2.7). The matrix is symmetric and has the form

$$\mathbf{C} = \begin{bmatrix} c_{11} & c_{12} & \cdots & c_{1p} \\ c_{21} & c_{22} & \cdots & c_{2p} \\ \vdots & \vdots & & \vdots \\ c_{p1} & c_{p2} & \cdots & c_{pp} \end{bmatrix},$$

where the diagonal element c_{ii} is the variance of X_i and c_{ij} is the covariance of variables X_i and X_j.

The variances of the principal components are the eigenvalues of the matrix $\mathbf{C}$. There are p of these, some of which may be zero. Negative eigenvalues are not possible for a covariance matrix. Assuming that the eigenvalues are ordered as $\lambda_1 \geqslant \lambda_2 \geqslant \cdots \geqslant \lambda_p \geqslant 0$, then λ_i corresponds to the ith principal component

$$Z_i = a_{i1}x_1 + a_{i2}X_2 + \cdots + a_{ip}X_p.$$

In particular, var(Z_i) = λ_i and the constants $a_{i1}, a_{i2}, \ldots, a_{ip}$ are the elements of the corresponding eigenvector, scaled so that $a_{i1}^2 + a_{i2}^2 + \cdots + a_{ip}^2 = 1$.

An important property of the eigenvalues is that they add up to the sum of the diagonal elements (the trace) of $\mathbf{C}$. That is

$$\lambda_1 + \lambda_2 + \cdots + \lambda_p = c_{11} + c_{22} + \cdots + c_{pp}.$$

As c_{ii} is the variance of X_i and λ_i is the variance of Z_i, this means that the sum of the variances of the principal components is equal to the sum of the variances of the original variables. Therefore, in a sense, the principal components account for all of the variation in the original data.

In order to avoid one variable having an undue influence on the principal components it is usual to code the variables $X_1, X_2, \ldots, X_p$ to have means of zero and variances of one at the start of an analysis. The matrix $\mathbf{C}$ then takes the form

$$\mathbf{C} = \begin{bmatrix} 1 & c_{12} & \cdots & c_{1p} \\ c_{21} & 1 & \cdots & c_{2p} \\ \vdots & \vdots & & \vdots \\ c_{p1} & c_{p2} & \cdots & 1 \end{bmatrix}$$

where $c_{ij} = c_{ji}$ is the correlation between X_i and X_j. In other words, the principal components analysis is carried out on the correlation matrix. In that case, the sum of the diagonal terms, and hence the sum of the eigenvalues, is equal to p, the number of variables.

The steps in a principal components analysis can now be stated:

1. Start by coding the variables $X_1, X_2, \ldots, X_p$ to have zero means and unit variances. This is usual, but is omitted in some cases.
2. Calculate the covariance matrix $\mathbf{C}$. This is a correlation matrix if step 1 has been done.
3. Find the eigenvalues $\lambda_1, \lambda_2, \ldots, \lambda_p$ and the corresponding eigenvectors $\mathbf{a}_1, \mathbf{a}_2, \ldots, \mathbf{a}_p$. The coefficients of the ith principal component are then given by $\mathbf{a}_i$ while λ_i is its variance.
4. Discard any components that only account for a small proportion of the variation in the data. For example, starting with 20 variables it might be found that the first three components account for 90% of the total variance. On this basis the other 17 components may reasonably be ignored.

Example 6.1 Body measurements of female sparrows

Some mention has already been made of what happens when a principal components analysis is carried out on the data on five body measurements of 49 female sparrows (Table 1.1). It is now worth while to consider the example in more detail.

It is appropriate to begin with step 1 of the four parts of the analysis that have just been described. Standardization of the measurements ensures that they all have equal weight in the analysis. Omitting standardization would mean that the variables X_1 and X_2, which vary most over the 49 birds, would tend to dominate the principal components.

The covariance matrix for the standardized variables is the correlation matrix. This has already been given in lower triangular form in Table 6.1. The eigenvalues of this matrix are found to be 3.616, 0.532, 0.386, 0.302 and 0.164. These add to 5.000, the sum of the diagonal terms in the correlation matrix. The corresponding eigenvectors are shown in Table 6.3, standardized so that the sum of the squares of the coefficients is unity for each one of them. These eigenvectors then provide the coefficients of the principal components.

The eigenvalue for a principal component indicates the variance that it accounts for out of the total variances of 5.000. Thus the first principal component accounts for $(3.616/5.000) \, 100\% = 72.3\%$, the second for 10.6%, the third for 7.7%, the fourth for 6.0%, and the fifth for 3.3%. Clearly the first component is far more important than the others.

Table 6.3 The eigenvalues and eigenvectors of the correlation matrix for five measurements on 49 female sparrows. The eigenvalues are the variances of the principal components. The eigenvectors give the coefficients of the standardized variables

		Eigenvector, coefficient of				
Component	Eigenvalue	X_1	X_2	X_3	X_4	X_5
1	3.616	0.452	0.462	0.451	0.471	0.398
2	0.532	−0.051	0.300	0.325	0.185	−0.877
3	0.386	0.691	0.341	−0.455	−0.411	−0.179
4	0.302	−0.420	0.548	−0.606	0.388	0.069
5	0.165	0.374	−0.530	−0.343	0.652	−0.192

Another way of looking at the relative importance of principal components is in terms of their variance in comparison to the variance of the original variables. After standardization the original variables all have variances of 1.0. The first principal component therefore has a variance of 3.616 original variables. However, the second principal component has a variance of only 0.532 of that of one of the original variables. The other principal components account for even less variation.

The first principal component is

$$Z_1 = 0.452X_1 + 0.462X_2 + 0.451X_3 + 0.471X_4 + 0.398X_5,$$

where X_1 to X_5 are standardized variables. This is an index of the size of the sparrows. It seems therefore that about 72.3% of the variation in the data are related to size differences.

The second principal component is

$$Z_2 = -0.051X_1 + 0.300X_2 + 0.325X_3 + 0.185X_4 - 0.877X_5.$$

This appears to be a contrast between variables X_2 (alar extent), X_3 (length of beak and head), and X_4 (length of humerus) on the one hand, and variable X_5 (length of the keel of the sternum) on the other. That is to say, Z_2 will be high if X_2, X_3 and X_4 are high but X_5 is low. On the other hand, Z_2 will be low if X_2, X_3 and X_4 are low but X_5 is high. Hence Z_2 represents a shape difference between the sparrows. The low coefficient of X_1 (total length) means that the value of this variable does not affect Z_2 very much. The principal components Z_3, Z_4 and Z_5 can be interpreted in a similar way. They represent other aspects of shape differences.

The values of the principal components may be useful for further analyses. They are calculated in the obvious way from the standardized variables. Thus for the first bird the original variable values are $x_1 = 156$, $x_2 = 245, x_3 = 31.6, x_4 = 18.5$ and $x_5 = 20.5$. These standardize to $x_1 = (156 - 157.980)/3.654 = -0.542$, $x_2 = (245 - 241.327)/5.068 = 0.725$, $x_3 = (31.6 - 31.459)/0.795 = 0.177$, $x_4 = (18.5 - 18.469)/0.564 = 0.055$, and $x_5 = (20.5 - 20.827)/0.991 = -0.330$, where in each case the variable mean for the 49 birds has been subtracted and a division has been made by the variable standard deviation for the 49 birds. The value of the first principal component for the first bird is therefore

$$Z_1 = 0.452 \times (-0.542) + 0.462 \times 0.725 + 0.451 \times 0.177$$
$$+ 0.471 \times 0.055 + 0.398 \times (-0.330)$$
$$= 0.064.$$

The second principal component for the bird is

$$Z_2 = -0.051 \times (-0.542) + 0.300 \times 0.725 + 0.325 \times 0.177$$
$$+ 0.185 \times 0.055 - 0.877 \times (-0.330)$$
$$= 0.602.$$

The other principal components can be calculated in a similar way.

The birds being considered were picked up after a severe storm. The first 21 of them recovered while the other 28 died. A question of some interest is therefore whether the survivors and non-survivors show any differences. It has been shown in Example 4.1 that there is no evidence of any differences in mean values. However, in Example 4.2 it has been shown that the survivors seem to have been less variable than the non-survivors. The situation can now be considered in terms of principal components.

The means and standard deviations of the five principal components are shown in Table 6.4 separately for survivors and non-survivors. None of the mean differences between survivors and non-survivors is significant on a t test and none of the standard

Table 6.4 Comparison between survivors and non-survivors in terms of means and standard deviations of principal components

Principal component	Mean		Standard deviation	
	Survivors	Non-survivors	Survivors	Non-survivors
1	−0.100	0.075	1.506	2.176
2	0.004	−0.003	0.684	0.776
3	−0.140	0.105	0.522	0.677
4	0.073	−0.055	0.563	0.543
5	0.023	−0.017	0.411	0.408

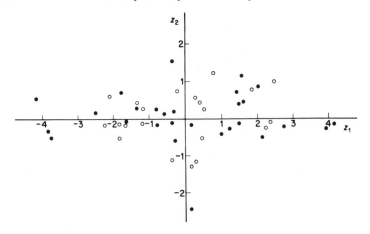

Figure 6.1 Plot of 49 female sparrows against values for the first two principal components, Z_1 and Z_2. (Open circles indicate survivors, closed circles indicate non-survivors.)

deviation differences is significant on an F test. However, Levene's test on deviations from medians (described in Chapter 4) gives a significant difference (just) between the variation of principal component 1 for survivors and non-survivors on a one-sided test at the 5% level. The assumption for the one-sided test is that, if anything, non-survivors were more variable than survivors. The variation is not significantly different for survivors and non-survivors with Levene's test on the other principal components. As principal component 1 measures overall size, it seems that stabilizing selection may have acted against very large and very small birds.

Figure 6.1 shows a plot of the values of the 49 birds for the first two principal components, which between them account for 82.9% of the variation in the data. The figure shows quite clearly how birds with extreme values for the first principal component failed to survive. Indeed, there is a suggestion that this was true for principal component 2 as well.

Example 6.2 Employment in European countries

As a second example of a principal components analysis, consider the data in Table 1.5 on the percentages of people employed in nine industry sectors in Europe in 1979. The correlation matrix for the

Table 6.5 The correlation matrix for percentages employed in nine industry groups in 26 countries in Europe, in lower diagonal form, calculated from Table 1.5

	AGR	MIN	MAN	PS	CON	SER	FIN	SPS	TC
Agriculture	1.000								
Mining	0.036	1.000							
Manufacturing	-0.671	0.445	1.000						
Power supplies	-0.400	0.406	0.385	1.000					
Construction	-0.538	-0.026	0.495	0.060	1.000				
Service industries	-0.737	-0.397	0.204	0.202	0.356	1.000			
Finance	-0.220	-0.443	-0.156	0.110	0.016	0.366	1.000		
Social & personal services	-0.747	-0.281	0.154	0.132	0.158	0.572	0.108	1.000	
Transport & communications	-0.565	0.157	0.351	0.375	0.388	0.188	-0.246	0.568	1.000

nine variables is shown in Table 6.5. Overall the values in this matrix are not particularly high, which indicates that several principal components will be required to account for the variation.

The eigenvalues of the correlation matrix, with percentages of the total of 9.000 in parentheses, are 3.487(38.7%), 2.130(23.6%), 1.099(12.2%), 0.995(11.1%), 0.543(6.0%), 0.383(4.2%), 0.226(2.5%), 0.137(1.5%) and 0.000(0%). The last eigenvalue is exactly zero because the sum of the nine variables being analysed is 100% before standardization. The eigenvector corresponding to this eigenvalue is precisely this sum which, of course, has a zero variance. If any linear combination of the original variables in a principal components analysis is constant then this must of necessity result in a zero eigenvalue.

This example is not as straightforward as the previous one. The first principal component only accounts for about 40% of the variation in the data and four components are needed to account for 86% of the variation. It is a matter of judgement as to how many components are important. It can be argued that only the first two should be considered because these are the only ones with eigenvalues much more than 1.000. On the other hand, the first four components all have eigenvalues substantially larger than the last five components so perhaps the first four should all be considered. To some extent the choice of the number of components that are important will depend on the use that is going to be made of them. For the present example it will be assumed that a small number of indices are required in order to present the main aspects of differences between the countries. For simplicity only the first two components will be examined further. Between them they account for about 62% of the variation.

The first component is

$$Z_1 = 0.52(\text{AGR}) + 0.00(\text{MIN}) - 0.35(\text{MAN}) - 0.26(\text{PS})$$
$$- 0.33(\text{CON}) - 0.38(\text{SER}) - 0.07(\text{FIN}) - 0.39(\text{SPS})$$
$$- 0.37(\text{TC}),$$

where the abbreviations for variables are stated in Table 1.5. As the analysis has been done on the correlation matrix, the variables in this equation are the original percentages after they have each been standardized to have a mean of zero and a standard deviation of one. From the coefficients of Z_1 it can be seen that it is primarily a contrast between numbers engaged in agriculture (AGR) and numbers engaged in manufacturing (MAN), power supplies (PS), construction (CON), service industries (SER), social and personal services (SPS) and transport and communications (TC). In making this interpretation the variables with coefficients close to zero are ignored because they will not affect the value of Z_1 greatly.

The second component is

$$Z_2 = 0.05(AGR) + 0.62(MIN) + 0.36(MAN) + 0.26(PS)$$
$$+ 0.05(CON) - 0.35(SER) - 0.45(FIN) - 0.22(SPS)$$
$$+ 0.20(TC),$$

which primarily contrasts numbers in mining (MIN) and manufacturing (MAN) with numbers in service industries (SER) and finance (FIN).

Figure 6.2 shows a plot of the 26 countries against their values for Z_1 and Z_2. The picture is certainly rather meaningful in terms of what is known about the countries as they were in 1979. Most of the Western democracies are grouped with low values of Z_1 and Z_2. Ireland, Portugal, Spain and Greece have higher values of Z_1. Turkey and Yugoslavia stand out as being very high on Z_1. The communist countries other than Yugoslavia are grouped together with high values for Z_2.

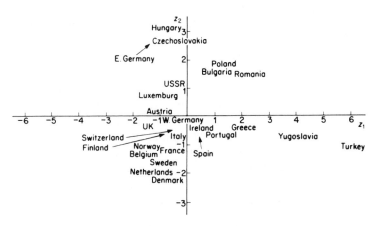

Figure 6.2 European countries plotted against the first two principal components, Z_1 and Z_2, for employment variables.

6.3 Computational methods and computer programs

Many standard statistical packages will carry out a principal components analysis because it is one of the commonest types of multivariate analysis in use. When the analysis is not mentioned as an option it may still be possible to do the required calculations as a special type of factor analysis (as explained in Chapter 7). It is, of course, also always possible to find the eigenvalues and eigenvectors of a covariance or correlation matrix using a suitable computer program, where that represents the most complicated part of principal components analysis.

6.4 Further reading

Principal components analysis is covered in almost all texts on multivariate analysis, and in great detail by Jolliffe (1986) and Jackson (1991). Social scientists may also find the shorter monograph by Dunteman (1989) to be helpful.

Exercises

1. Table 6.6 shows six measurements on each of 25 pottery goblets excavated from prehistoric sites in Thailand, with Fig. 6.3 illustrating the typical shape and the nature of the measurements. The main question of interest for these data concerns similarities and differences between the goblets, with obvious questions being:
   (a) Is it possible to display the data graphically to show how the goblets are related and, if so, are there any obvious groupings of similar goblets?
   (b) Are there any goblets that are particularly unusual?
   Carry out a principal components analysis and see whether the values of the principal components help to answer these questions.
   One point that needs consideration with this exercise is the extent to which differences between goblets are due to shape differences rather than size differences. It may well be considered that two goblets that are almost the same shape but have very different sizes are really 'similar'. The problem of separating size and shape differences has generated a considerable scientific literature that will not be considered here. However, it can be noted that one way to remove the effects of size involves dividing

Table 6.6 Measurements (in cm) taken on 25 prehistoric goblets from Thailand. The variables are defined in Fig. 6.3. The data were kindly provided by Professor C.F.W. Higham of the University of Otago

Goblet	X_1	X_2	X_3	X_4	X_5	X_6
1	13	21	23	14	7	8
2	14	14	24	19	5	9
3	19	23	24	20	6	12
4	17	18	16	16	11	8
5	19	20	16	16	10	7
6	12	20	24	17	6	9
7	12	19	22	16	6	10
8	12	22	25	15	7	7
9	11	15	17	11	6	5
10	11	13	14	11	7	4
11	12	20	25	18	5	12
12	13	21	23	15	9	8
13	12	15	19	12	5	6
14	13	22	26	17	7	10
15	14	22	26	15	7	9
16	14	19	20	17	5	10
17	15	16	15	15	9	7
18	19	21	20	16	9	10
19	12	20	26	16	7	10
20	17	20	27	18	6	14
21	13	20	27	17	6	9
22	9	9	10	7	4	3
23	8	8	7	5	2	2
24	9	9	8	4	2	2
25	12	19	27	18	5	12

the measurements for a goblet by the total height of the body of the goblet. Alternatively, the measurements on a goblet can be expressed as a proportion of the sum of all measurements on that goblet. These types of standardization of variables will clearly ensure that the data values are similar for two goblets with the same shape but different sizes.

2. Table 6.7 shows estimates of the average protein consumption from different food sources for the inhabitants of 25 European countries as published by Weber (1973). Use principal components

Table 6.7 Protein consumption (g per person per day) in 25 European countries

Country	Red meat	White meat	Eggs	Milk	Fish	Cereals	Starchy foods	Pulses, nuts and oilseeds	Fruits and Vegetables	Total
Albania	10	1	1	9	0	42	1	6	2	72
Austria	9	14	4	20	2	28	4	1	4	86
Belgium	14	9	4	18	5	27	6	2	4	89
Bulgaria	8	6	2	8	1	57	1	4	4	91
Czechoslovakia	10	11	3	13	2	34	5	1	4	83
Denmark	11	11	4	25	10	22	5	1	2	91
E. Germany	8	12	4	11	5	25	7	1	4	77
Finland	10	5	3	34	6	26	5	1	1	91
France	18	10	3	20	6	28	5	2	7	99
Greece	10	3	3	18	6	42	2	8	7	99
Hungary	5	12	3	10	0	40	4	5	4	83
Ireland	14	10	5	26	2	24	6	2	3	92
Italy	9	5	3	14	3	37	2	4	7	84
Netherlands	10	14	4	23	3	22	4	2	4	86
Norway	9	5	3	23	10	23	5	2	3	83
Poland	7	10	3	19	3	36	6	2	7	93
Portugal	6	4	1	5	14	27	6	5	8	76
Romania	6	6	2	11	1	50	3	5	3	87
Spain	7	3	3	9	7	29	6	6	7	77
Sweden	10	8	4	25	8	20	4	1	2	82
Switzerland	13	10	3	24	2	26	3	2	5	88
UK	17	6	5	21	4	24	5	3	3	88
USSR	9	5	2	17	3	44	6	3	3	92
W. Germany	11	13	4	19	3	19	5	2	4	80
Yugoslavia	4	5	1	10	1	56	3	6	3	89

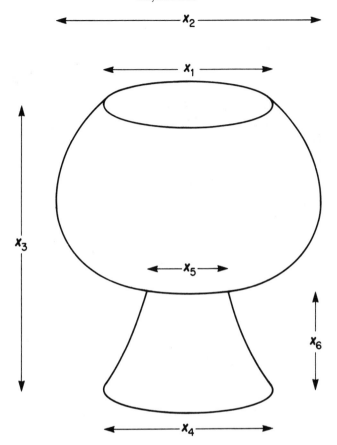

Figure 6.3 Measurements made on pottery goblets from Thailand.

analysis to investigate the relationships between the countries on the basis of these variables.

References

Dunteman, G.H. (1989) *Principal Components Analysis.* Sage Publications, Newbury Park, California.
Hotelling, H. (1933) Analysis of a complex of statistical variables into principal components. *Journal of Educational Psychology* **24**, 417–41 and 498–520.

Jackson, J.E. (1991) *A User's Guide to Principal Components.* Wiley, New York.

Jolliffe, I.T. (1986) *Principal Component Analysis.* Springer-Verlag, New York.

Pearson, K. (1901) On lines and planes of closest fit to a system of points in space. *Philosophical Magazine* **2**, 557–72.

Weber, A. (1973) *Agrarpolitik im Spannungsfeld der Internationalen Ernährungspolitik.* Institut für Agrapolitik und Marktlehre, Kiel.

CHAPTER SEVEN

Factor analysis

7.1 The factor analysis model

Factor analysis has similar aims to principal components analysis. The basic idea is still that it may be possible to describe a set of p variables $X_1, X_2, \ldots, X_p$ in terms of a smaller number of indices or factors, and hence elucidate the relationship between these variables. There is, however, one important difference: principal components analysis is not based on any particular statistical model, but factor analysis is based on a rather special model.

The early development of factor analysis was due to Charles Spearman. He studied the correlations between test scores of various types and noted that many observed correlations could be accounted for by a simple model for the scores (Spearman, 1904). For example, in one case he obtained the following matrix of correlations for boys in a preparatory school for their scores on tests in Classics (C), French (F), English (E), Mathematics (M), Discrimination of pitch (D), and Music (Mu):

$$
\begin{array}{c}
\\ C \\ F \\ E \\ M \\ D \\ Mu
\end{array}
\begin{bmatrix}
C & F & E & M & D & Mu \\
1.00 & 0.83 & 0.78 & 0.70 & 0.66 & 0.63 \\
0.83 & 1.00 & 0.67 & 0.67 & 0.65 & 0.57 \\
0.78 & 0.67 & 1.00 & 0.64 & 0.54 & 0.51 \\
0.70 & 0.67 & 0.64 & 1.00 & 0.45 & 0.51 \\
0.66 & 0.65 & 0.54 & 0.45 & 1.00 & 0.40 \\
0.63 & 0.57 & 0.51 & 0.51 & 0.40 & 1.00
\end{bmatrix}
$$

He noted that this matrix has the interesting property that any two rows are almost proportional if the diagonals are ignored. Thus for rows C and E there are ratios:

$$
\frac{0.83}{0.67} \simeq \frac{0.70}{0.64} \simeq \frac{0.66}{0.54} \simeq \frac{0.63}{0.51} \simeq 1.2.
$$

Spearman proposed the idea that the six test scores are all of the form

$$X_i = a_i F + e_i,$$

where X_i is the ith standardized score with a mean of zero and a standard deviation of one, a_i is a constant, F is a 'factor' value, which has mean of zero and standard deviation of one for individuals as a whole, and e_i is the part of X_i that is specific to the ith test only. He showed that a constant ratio between rows of a correlation matrix follows as a consequence of these assumptions and that therefore this is a plausible model for the data.

Apart from the constant correlation ratios it also follows that the variance of X_i is given by

$$\begin{aligned}
\text{var}(X_i) &= \text{var}(a_i F + e_i) \\
&= \text{var}(a_i F) + \text{var}(e_i) \\
&= a_i^2 \text{var}(F) + \text{var}(e_i) \\
&= a_i^2 + \text{var}(e_i),
\end{aligned}$$

because a_i is a constant, F and e_i are independent, and the variance of F is assumed to be unity. But $\text{var}(X_i)$ is also unity, so that

$$1 = a_i^2 + \text{var}(e_i).$$

Hence the constant a_i, which is called the *factor loading*, is such that its square is the proportion of the variance of X_i that is accounted for by the factor.

On the basis of his work Spearman formulated his two-factor theory of mental tests: each test result is made up of two parts, one that is common to all tests ('general intelligence'), and another that is specific to the test. Later this theory was modified to allow each test result to consist of a part due to several common factors plus a part specific to the test. This gives the general factor analysis model

$$X_i = a_{i1} F_1 + a_{i2} F_2 + \cdots + a_{im} F_m + e_i,$$

where X_i is the ith test score with mean zero and unit variance; $a_{i1}, a_{i2}, \ldots, a_{im}$ are the *factor loadings* for the ith test; $F_1, F_2, \ldots, F_m$ are m uncorrelated *common factors*, each with mean zero and unit variance; and e_i is a factor specific only to the ith test, which is uncorrelated with any of the common factors and has mean zero.

With this model

$$\text{var}(X_i) = 1 = a_{i1}^2 \text{var}(F_1) + a_{i2}^2 \text{var}(F_2) + \cdots + a_{im}^2 \text{var}(F_m) + \text{var}(e_i)$$
$$= a_{i1}^2 + a_{i2}^2 + \cdots + a_{im}^2 + \text{var}(e_i),$$

where $a_{i1}^2 + a_{i2}^2 + \cdots + a_{im}^2$ is called the *communality* of X_i (the part of its variance that is related to the common factors) while $\text{var}(e_i)$ is called the *specificity* of X_i (the part of its variance that is unrelated to the common factors). It can also be established that the correlation between X_i and X_j is

$$r_{ij} = a_{i1}a_{j1} + a_{i2}a_{j2} + \cdots + a_{im}a_{jm}.$$

Hence two test scores can only be highly correlated if they have high loadings on the same factors. Furthermore, $-1 \leqslant a_{ij} \leqslant +1$ as the communality cannot exceed one.

7.2 Procedure for a factor analysis

The data for a factor analysis have the same form as for a principal components analysis. That is, there are p variables with values for these n individuals, as shown in Table 6.2.

There are three stages to a factor analysis. To begin with, provisional factor loadings a_{ij} are determined. One way to do this is to do a principal components analysis and neglect all of the principal components after the first m, which are themselves taken to be the m factors. The factors found in this way are then uncorrelated with each other and are also uncorrelated with the specific factors. However, the specific factors are not uncorrelated with each other, which means that one of the assumptions of the factor analysis model does not hold. However, this may not matter much providing that the communalities are high.

Whatever way the provisional factor loadings are determined, it is possible to show they are not unique. If $F_1, F_2, \ldots, F_m$ are the provisional factors, then linear combinations of these of the form

$$F'_1 = d_{11}F_1 + d_{12}F_2 + \cdots + d_{1m}F_m$$
$$F'_2 = d_{21}F_1 + d_{22}F_2 + \cdots + d_{2m}F_m$$
$$\vdots$$
$$F'_m = d_{m1}F_1 + d_{m2}F_2 + \cdots + d_{mm}F_m$$

can be constructed that are uncorrelated and 'explain' the data just as well. There are an infinite number of alternative solutions for the factor analysis model, and this leads to the second stage in the analysis, which is called *factor rotation*. Thus the provisional factors are transformed in order to find new factors that are easier to interpret. To 'rotate' in this context means essentially to choose the d_{ij} values in the above equations.

The last stage of an analysis involves calculating the *factor scores*. These are the values of the factors $F_1, F_2, \ldots, F_m$ for each of the individuals.

Generally the number of factors (m) is up to the factor analyst, although it may sometimes be suggested by the nature of the data. When a principal components analysis is used to find a provisional solution, a rough 'rule of thumb' is to choose m equal to the number of eigenvalues greater than unity for the correlation matrix of the test scores. The logic here is the same as was explained in the previous chapter: a factor associated with an eigenvalue of less than unity 'explains' less variation in the overall data than one of the original test scores. In general, increasing m will increase the communalities of variables. However, communalities are not changed by factor rotation.

Factor rotation can be *orthogonal* or *oblique*. With orthogonal rotation the new factors are uncorrelated, like the old factors. With oblique rotation the new factors are correlated. Whichever type of rotation is used, it is desirable that the factor loadings for the new factors should be either close to zero or very different from zero. A near zero a_{ij} means that X_i is not strongly related to the factor F_j. A large (positive or negative) value of a_{ij} means that X_i is determined by F_j to a large extent. If each test score is strongly related to some factors, but not related to the others, then this makes the factors easier to identify than would otherwise be the case.

One method of orthogonal factor rotation that is often used is called *varimax rotation*. This is based on the assumption that the interpretability of factor j can be measured by the variance of the square of its factor loadings, i.e. the variance of $a_{1j}^2, a_{2j}^2, \ldots, a_{pj}^2$. If this variance is large then the a_{ij}^2 values tend to be either close to zero or close to unity. Varimax rotation therefore maximizes the sum of these variances for all the factors. H.F. Kaiser first suggested this approach. Later he modified it slightly by normalizing the factor loadings before maximizing the variances of their squares, since this appears to give improved results (Kaiser, 1958). Varimax rotation can therefore be carried out with or without Kaiser normalization. Numerous other methods or orthogonal rotation have been proposed. However, varimax is recommended as the standard approach.

Sometimes factor analysts are prepared to give up the idea of the factors being uncorrelated in order that the factor loadings should be as simple as possible. An oblique rotation may then give a better solution than an orthogonal one. Again, there are numerous methods available to do the oblique rotation.

Various methods have also been suggested for calculating the factor scores for individuals. A method for use with factor analysis based on principal components is described in the next section. Two other more general methods are estimation by regression and Bartlett's method. See Harman (1976, Chapter 16) for more details.

7.3 Principal components factor analysis

It has been remarked above that one way to do a factor analysis is to begin with a principal components analysis and use the first few principal components as unrotated factors. This has the virtue of simplicity although as the specific factors $e_1, e_2, \ldots, e_p$ are correlated the factor analysis model is not quite correct. Experienced factor analysts often do a principal components factor analysis first and then use other approaches.

The method for finding the unrotated factors is as follows. With p variables there will be the same number of principal components, these being of the form

$$
\left.
\begin{aligned}
Z_1 &= b_{11}X_1 + b_{12}X_2 + \cdots + b_{1p}X_p \\
Z_2 &= b_{21}X_1 + b_{22}X_2 + \cdots + b_{2p}X_p \\
&\;\vdots \\
Z_p &= b_{p1}X_1 + b_{p2}X_2 + \cdots + b_{pp}X_p
\end{aligned}
\right\}
\qquad (7.1)
$$

where the b_{ij} values are given by the eigenvectors of the correlation matrix. This transformation from X values to Z values is orthogonal, so that the inverse relationship is simply

$$X_1 = b_{11}Z_1 + b_{21}Z_2 + \cdots + b_{p1}Z_p$$
$$X_2 = b_{12}Z_1 + b_{22}Z_2 + \cdots + b_{p2}Z_p$$
$$\vdots$$
$$X_p = b_{1p}Z_1 + b_{2p}Z_2 + \cdots + b_{pp}Z_p.$$

For a factor analysis only m of the principal components are retained, so the last equations become

$$X_1 = b_{11}Z_1 + b_{21}Z_2 + \cdots + b_{m1}Z_m + e_1$$
$$X_2 = b_{12}Z_1 + b_{22}Z_2 + \cdots + b_{m2}Z_m + e_2$$
$$\vdots$$
$$X_p = b_{1p}Z_1 + b_{2p}Z_2 + \cdots + b_{mp}Z_m + e_p$$

where e_i is a linear combination of the principal components, Z_{m+1} to Z_p. All that needs to be done now is to scale the principal components $Z_1, Z_2, \ldots, Z_m$ to have unit variances and hence make them into proper factors. To do this Z_i must be divided by its standard deviation, which is $\sqrt{\lambda_i}$, the square root of the corresponding eigenvalue in the correlation matrix. The equations then become

$$X_1 = \sqrt{\lambda_1}b_{11}F_1 + \sqrt{\lambda_2}b_{21}F_2 + \cdots + \sqrt{\lambda_m}b_{m1}Z_m + e_1$$
$$X_2 = \sqrt{\lambda_1}b_{12}F_1 + \sqrt{\lambda_2}b_{22}F_2 + \cdots + \sqrt{\lambda_m}b_{m2}Z_m + e_2$$
$$\vdots$$
$$X_p = \sqrt{\lambda_1}b_{1p}F_1 + \sqrt{\lambda_2}b_{2p}F_2 + \cdots + \sqrt{\lambda_m}b_{mp}Z_m + e_p$$

where $F_i = Z_i/\sqrt{\lambda_i}$. The unrotated factor model is then

$$\left. \begin{array}{l} X_1 = a_{11}F_1 + a_{12}F_2 + \cdots + a_{1m}F_m + e_1 \\ X_2 = a_{21}F_1 + a_{22}F_2 + \cdots + a_{2m}F_m + e_2 \\ \vdots \\ X_p = a_{p1}F_1 + a_{p2}F_2 + \cdots + a_{pm}F_m + e_p \end{array} \right\} \tag{7.2}$$

where $a_{ij} = \sqrt{\lambda_j}b_{ji}$.

After rotation (say to varimax loadings) a new solution has the form

$$
\left.
\begin{aligned}
X_1 &= g_{11}F_1^* + g_{12}F_2^* + \cdots + g_{1m}F_m^* + e_1 \\
X_2 &= g_{21}F_1^* + g_{22}F_2^* + \cdots + g_{2m}F_m^* + e_2 \\
&\;\vdots \\
X_p &= g_{p1}F_1^* + g_{p2}F_2^* + \cdots + g_{pm}F_m^* + e_p
\end{aligned}
\right\}
\tag{7.3}
$$

where F_i^* represents the new ith factor. The original factors F_i can be expressed exactly as linear combinations of the X variables using equations (7.1). The rotated factors can also still be expressed exactly as linear combinations of the X variables, the relationship being given in matrix form as

$$
\mathbf{F}^* = (\mathbf{G'\, G})^{-1} \mathbf{G'\, X}
\tag{7.4}
$$

(Harman, 1976, p. 367), where $(\mathbf{F}^*)' = (F_1^*, F_2^*, \ldots, F_m^*)$, $\mathbf{X}' = (X_1, X_2, \ldots, X_p)$, and $\mathbf{G}$ is the $p \times m$ matrix of factor loadings given in equation (7.3).

7.4 Using a factor analysis program to do principal components analysis

Because many computer programs for factor analysis allow the option of using principal components as initial factors, it is possible to use the programs to do principal components analysis. All that has to be done is to extract the same number of factors as variables and not do any rotation. The factor loadings will then be as given by equation (7.2) with $m = p$ and $e_1 = e_2 = \cdots = e_m = 0$. The principal components are given by equations (7.1) with $b_{ij} = a_{ji}/\sqrt{\lambda_i}$, where λ_i is the ith eigenvalue.

Example 7.1 Employment in European countries

In Example 6.2 a principal components analysis was carried out on the data on the percentages of people employed in nine industry groups in 26 countries in Europe in 1979 (Table 1.5). It is of some interest to continue the examination of these data using a factor analysis model.

Table 7.1 Eigenvalues and vectors for the European employment data

		Eigenvector, coefficient of								
	Eigenvalue	X_1	X_2	X_3	X_4	X_5	X_6	X_7	X_8	X_9
1	3.487	0.524	0.001	-0.348	-0.256	-0.325	-0.379	-0.074	-0.387	-0.367
2	2.130	0.054	0.618	0.355	0.261	0.051	-0.350	-0.454	-0.222	0.203
3	1.099	-0.049	0.201	0.151	0.561	-0.153	0.115	0.587	-0.312	-0.378
4	0.995	0.029	0.064	-0.346	0.393	-0.668	-0.050	-0.052	0.412	0.314
5	0.543	0.213	-0.164	-0.385	0.295	0.472	-0.283	0.280	-0.220	0.513
6	0.383	-0.153	0.101	0.289	-0.357	-0.130	-0.615	0.526	0.263	0.124
7	0.226	0.021	-0.726	0.479	0.256	-0.211	0.229	-0.188	-0.191	0.068
8	0.137	0.008	0.088	0.126	-0.341	0.356	0.388	0.174	-0.506	0.545
9	0	-0.806	-0.049	-0.366	-0.019	-0.083	-0.238	-0.145	-0.351	-0.072

The correlation matrix for the nine percentage variables is given in Table 6.5. The eigenvalues and eigenvectors are shown in Table 7.1. There are three eigenvalues greater than unity so the 'rule of thumb' suggests that three factors should be considered. However, the fourth eigenvalue is almost equal to the third so that either two or four factors can also reasonably be allowed. To begin with, the four-factor solution will be considered.

The eigenvectors in Table 7.1 give the coefficients b_{ij} of the set of equations (7.1). These are changed into factor loadings for four factors as indicated in equations (7.2) to give the factor model:

$$X_1 = \quad \underline{0.98}F_1 + 0.08F_2 - 0.05F_3 + 0.03F_4 \qquad (0.97)$$

$$X_2 = \quad 0.00F_1 + \underline{0.90}F_2 + 0.21F_3 + 0.06F_4 \qquad (0.86)$$

$$X_3 = -\underline{0.65}F_1 + \underline{0.52}F_2 + 0.16F_3 - 0.35F_4 \qquad (0.83)$$

$$X_4 = -0.48F_1 + 0.38F_2 + \underline{0.59}F_3 + 0.39F_4 \qquad (0.87)$$

$$X_5 = -\underline{0.61}F_1 + 0.08F_2 - 0.16F_3 - \underline{0.67}F_4 \qquad (0.84)$$

$$X_6 = -\underline{0.71}F_1 - \underline{0.51}F_2 + 0.12F_3 - 0.05F_4 \qquad (0.78)$$

$$X_7 = -0.14F_1 - \underline{0.66}F_2 + \underline{0.62}F_3 - 0.05F_4 \qquad (0.84)$$

$$X_8 = -\underline{0.72}F_1 - 0.32F_2 - 0.33F_3 + 0.41F_4 \qquad (0.90)$$

$$X_9 = -\underline{0.69}F_1 + 0.30F_2 - 0.39F_3 + 0.31F_4 \qquad (0.81)$$

The values in parentheses are the communalities. For example, the communality for variable X_1 is $0.98^2 + 0.08^2 + (-0.05)^2 + 0.03^2 = 0.97$, apart from rounding errors. It can be seen that the communalities are fairly high. That is to say, most of the variance for the variables X_1 to X_9 is accounted for by the four common factors.

Factor loadings that are greater than 0.50 (ignoring the sign) are underlined in the above equations. These large and moderate loadings indicate how the variables are related to the factors. It can be seen that X_1 is almost entirely accounted for by factor 1 alone, X_2 is accounted for mainly by factor 2, X_3 is accounted for by factor 1 and factor 2, etc. An undesirable property of this choice of factors is that four of the nine X variables (X_3, X_5, X_6, X_7) are related strongly to two of the factors. This suggests that a rotation may provide simpler factors.

A varimax rotation with Kaiser normalization was carried out.

This produced the following model:

$$X_1 = \underline{0.68}F_1 - 0.27F_2 - 0.31F_3 + \underline{0.57}F_4$$

$$X_2 = 0.22F_1 + \underline{0.70}F_2 - \underline{0.55}F_3 - 0.13F_4$$

$$X_3 = -0.13F_1 + 0.49F_2 - 0.12F_3 - \underline{0.75}F_4$$

$$X_4 = -0.23F_1 + \underline{0.89}F_2 + 0.16F_3 - 0.02F_4$$

$$X_5 = -0.16F_1 - 0.11F_2 + 0.03F_3 - \underline{0.90}F_4$$

$$X_6 = -\underline{0.53}F_1 - 0.03F_2 + \underline{0.62}F_3 - 0.33F_4$$

$$X_7 = -0.07F_1 + 0.03F_2 + \underline{0.91}F_3 + 0.05F_4$$

$$X_8 = -\underline{0.93}F_1 - 0.05F_2 + 0.17F_3 - 0.04F_4$$

$$X_9 = -\underline{0.77}F_1 + 0.23F_2 - 0.33F_3 - 0.23F_4$$

The communalities are unchanged (apart from rounding errors) and the factors are still uncorrelated. However, this is a slightly better solution than the previous one as only X_1, X_2 and X_6 are now appreciably dependent on more than one factor.

At this stage it is usual to try to put labels on factors. It is fair to say that this often requires a degree of inventiveness and imagination! In the present case it is not too difficult.

Factor 1 has a high positive loading for X_1 (agriculture) and high negative loadings for X_6 (service industries), X_8 (social and personal services) and X_9 (transport and communications). It therefore measures the extent to which people are employed in agriculture rather than services, and communications. It can be labelled 'emphasis on agriculture and a lack of service industries'.

Factor 2 has high positive loadings for X_2 (mining) and X_4 (power supplies). This can be labelled 'emphasis on mining and power supplies'.

Factor 3 has high positive loadings on X_6 (service industries) and X_7 (finance) and a high negative loading on X_2 (mining). This can be labelled 'emphasis on financial and service industries rather than mining'.

Finally, factor 4 has high negative loadings on X_3 (manufacturing) and X_5 (construction) and a high positive loading on X_1 (agriculture). 'Lack of industrialization' seems to be a fair label in this case.

The **G** matrix of equations (7.3) and (7.4) is given by the factor

Table 7.2 Factor scores for 26 European countries

	Factor			
	1 *Agriculture and lack of service industries*	2 *Mining and power supplies*	3 *Financial and service industries and lack of mining*	4 *Lack of industrialization*
Belgium	−0.93	−0.04	0.86	−0.07
Denmark	−1.30	−1.10	0.59	0.44
France	0.02	−0.20	0.98	−0.43
W. Germany	−0.04	0.45	0.45	−0.32
Ireland	−0.32	0.37	0.35	0.82
Italy	0.08	−1.40	−0.07	−1.19
Luxembourg	0.37	0.59	0.18	−1.05
Netherlands	−0.90	−0.59	1.17	−0.24
UK	−0.85	1.23	0.95	0.59
Austria	0.06	0.83	0.68	−0.45
Finland	−0.92	0.47	0.62	0.74
Greece	0.56	−1.11	−0.56	0.42
Norway	−1.77	−0.67	−0.09	0.31
Portugal	0.40	−1.11	−0.07	−0.17
Spain	1.67	−0.64	0.93	−1.67
Sweden	−1.29	−0.38	0.61	0.67
Switzerland	0.67	−0.39	0.98	−1.62
Turkey	1.29	−1.57	−0.85	3.00
Bulgaria	0.26	−0.25	−1.39	−0.34
Czechoslovakia	0.30	1.18	−1.19	−0.63
E. Germany	−0.62	1.70	−1.19	−0.44
Hungary	−0.12	2.37	−1.07	0.42
Poland	0.42	0.26	−1.41	0.06
Romania	1.55	−0.29	−1.11	−0.67
USSR	−0.99	−0.87	−2.06	−0.06
Yugoslavia	2.35	1.17	1.70	1.90

loadings above. For example, $g_{11} = 0.68$ and $g_{12} = -0.27$, to two decimal places. Carrying out the matrix multiplications and inversion of equation (7.4) produces the equations

$$F_1^* = \quad 0.176X_1 + 0.127X_2 + 0.147X_3 + \cdots - 0.430X_9$$
$$F_2^* = -0.082X_1 + 0.402X_2 + 0.176X_3 + \cdots + 0.014X_9$$
$$F_3^* = -0.122X_1 - 0.203X_2 - 0.025X_3 + \cdots - 0.304X_9$$

and

$$F_4^* = \quad 0.175X_1 - 0.031X_2 - 0.426X_3 + \cdots + 0.088X_9$$

for estimating factor scores from the data values (after the X variables have been standardized to have zero mean and unit standard deviations). The factor scores obtained from these equations are given in Table 7.2 for the 26 European countries.

From studying the factor scores it can be seen that factor 1 emphasizes the importance of agriculture rather than services and communications in Yugoslavia, Spain, and Romania. The values of factor 2 indicate that countries like Hungary and East Germany had large numbers of people employed in mining and power supplies, with the situation being reversed in countries like Turkey and Italy. Factor 3, on the other hand, is mainly indicating the difference between the former communist bloc and the other countries in terms of the numbers employed in finance and service industries. Finally, the values for factor 4 mainly indicate how different Turkey is from the other countries because of its lack of manufacturing and construction workers and relatively large number of agricultural workers.

Most factor analysts would probably continue their analysis of this set of data by trying models with fewer factors and different methods of factor extraction. However, sufficient has been said already to indicate the general approach so the example will be left at this point.

7.5 Options in computer programs

Computer programs for factor analysis often allow many different options, which is likely to be rather confusing for the novice in this area. For example, the factor analysis module in BMDP (Dixon, 1990) allows four different methods for the initial extraction of factors and eight different methods for rotating these factors (including no rotation). This gives 32 different types of factor analysis that can all be expected to give somewhat different results. There is a similar range of choices with the other large standard packages such as SAS (1985) and SPSS (1990)

There is also the question of the number of factors to extract. Most computer programs make an automatic choice, but this may or may not be acceptable. The possibility of trying different numbers of factors therefore increases the choices for an analysis even more.

On the whole it is probably best to avoid using too many options when first practising factor analysis. The use of principal components as initial factors with varimax rotation, as used in the example in this chapter, is a reasonable start with any set of data. The maximum likelihood method for extracting factors is a good approach in principle, and might also be tried if this is available in the computer program being used.

7.6 The value of factor analysis

Factor analysis is something of an art, and it is certainly not as objective as most statistical methods. For this reason, many statisticians are sceptical about its value. For example, Chatfield and Collins (1980, p. 89) list six problems with factor analysis and conclude that 'factor analysis should not be used in most practical situations'. Similarly, Seber (1984) notes as a result of simulation studies that even if the postulated factor model is correct then the chance of recovering it using available methods is not high.

On the othe hand, factor analysis is widely used to analyse data and, no doubt, will continue to be widely used in future. The reason for this is that users find the results useful for gaining insight into the structure of multivariate data. Therefore, if it is thought of as a purely descriptive tool, with limitations that are understood, then it must take its place as one of the important multivariate methods. What must be avoided is carrying out a factor analysis on a single small sample that cannot be replicated and then assuming that the factors obtained must represent underlying variables that exist in the real world.

7.7 Computational methods and computer programs

This chapter has stressed factor analysis based on using principal components for unrotated factors, followed by varimax rotation.

This method is widely available in computer programs, and is often the default option. There should therefore be little difficulty in obtaining suitable software if this approach is used. The use of alternative methods for factor extraction and rotation is likely to require one of the larger statistical packages mentioned in section 7.5. The calculations for Example 7.1 were carried out using SOLO (BMDP, 1989).

7.8 Futher reading

Factor analysis is discussed in many texts on multivariate analysis although, as noted earlier, the topic is often not presented enthusiastically (Chatfield and Collins, 1980; Seber, 1984). There are some specialist books available on factor analysis such as the one written by Harman (1976). None seem to be of recent origin.

Exercise

Using Example 7.1 as a model, carry out a factor analysis of the data in Table 6.7 on protein consumption from 10 different food sources for the inhabitants of 25 European countries. Identify the important factors underlying the observed variables and examine the relationships between the countries with respect to these factors.

References

BMDP (1989) *SOLO User's Guide, Version 3.1*. BMDP Statistical Software, 1440 Sepulveda Boulevard, Los Angeles, California 90025.

Chatfield, C. and Collins, A.J. (1980) *Introduction to Multivariate Analysis*. Chapman and Hall, London.

Dixon, W.J. (ed.) (1990) *BMDP Statistical Software Manual*. University of California Press, Berkeley.

Harman, H.H. (1976) *Modern Factor Analysis*, 3rd edn. University of Chicago Press, Chicago.

Kaiser, H.F. (1958) The varimax criterion for analytic rotation in factor analysis. *Psychometrika* **23**, 187–200.

SAS (1985) *SAS User's Guide: Statistics*. SAS Institute, Cary, North Carolina 27511.

Seber, G.A.F. (1984) *Multivariate Observations*. Wiley, New York.

Spearman, C. (1904) 'General intelligence', objectively determined and measured. *American Journal of Psychology* **15**, 201–93.

SPSS (1990) *SPSS Reference Guide*. SPSS Inc., 444 N. Michigan Avenue, Chicago, Illinois 60611.

CHAPTER EIGHT

Discriminant function analysis

8.1 The problem of separating groups

The problem that is addressed with discriminant function analysis is how well it is possible to separate two or more groups of individuals, given measurements for these individuals on several variables. For example, with the data in Table 1.1 on five body measurements of 21 surviving and 28 non-surviving sparrows it is interesting to consider whether it is possible to use the body measurements to separate survivors and non-survivors. Also, for the data shown in Table 1.2 on four dimensions of Egyptian skulls for samples from five time periods it is reasonable to consider whether the measurements can be used to 'age' the skulls.

In the general case there will be m random samples from different groups, of sizes $n_1, n_2, \ldots, n_m$, and values will be available for p variables $X_1, X_2, \ldots, X_p$ for each sample member. Thus the data for

Table 8.1 The form of data for a discriminant function analysis

Individual	X_1	X_2	$\cdots$	X_p	
1	x_{111}	x_{112}	$\cdots$	x_{11p}	
2	x_{211}	x_{212}	$\cdots$	x_{21p}	Group 1
$\vdots$	$\vdots$	$\vdots$	$\vdots$	$\vdots$	
n_1	n_{n_111}	x_{n_112}	$\cdots$	x_{n_11p}	
1	x_{121}	x_{122}	$\cdots$	x_{12p}	
2	x_{221}	x_{222}	$\cdots$	x_{22p}	Group 2
$\vdots$	$\vdots$	$\vdots$	$\vdots$	$\vdots$	
n_2	x_{n_221}	x_{n_222}	$\cdots$	x_{n_22p}	
$\vdots$	$\vdots$	$\vdots$	$\vdots$	$\vdots$	
1	x_{1m1}	x_{1m2}	$\cdots$	x_{1mp}	
2	x_{2m1}	x_{2m2}	$\cdots$	x_{2mp}	Group m
$\vdots$	$\vdots$	$\vdots$	$\vdots$	$\vdots$	
n_m	x_{n_mm1}	x_{n_mm2}	$\cdots$	x_{n_mmp}	

a discriminant function analysis takes the form shown in Table 8.1.

The data for a discriminant function analysis do not need to be standardized to have zero means and unit variances prior to the start of the analysis, as is usual with principal components and factor analysis. This is because the outcome of a discriminant function analysis is not affected in any important way by the scaling of individual variables.

8.2 Discrimination using Mahalanobis distances

One approach to discrimination is based on Mahalanobis distances, as defined in section 5.3. The mean vectors for the m samples can be regarded as estimates of the true mean vectors for the groups. The Mahalanobis distances of individuals to group centres can then be calculated and each individual can be allocated to the group that it is closest to. This may or may not be the group that the individual actually came from. The percentage of correct allocations is clearly an indication of how well groups can be separated using the available variables.

This procedure is more precisely defined as follows. Let $\bar{\mathbf{x}}_i' = (\bar{x}_{1i}, \bar{x}_{2i}, \ldots, \bar{x}_{pi})'$ denote the vector of mean values for the sample from the ith group, calculated using equations (2.1) and (2.5), and let $\mathbf{C}_i$ denote the covariance matrix for the same sample calculated using equations (2.2), (2.3) and (2.7). Also, let $\mathbf{C}$ denote the pooled sample covariance matrix determined using equation (5.6). Then the Mahalanobis distance from an observation $\mathbf{x}' = (x_1, x_2, \ldots, x_p)'$ to the centre of group i is estimated as

$$D_i^2 = (\mathbf{x} - \bar{\mathbf{x}}_i)'\mathbf{C}^{-1}(\mathbf{x} - \bar{\mathbf{x}}_i)$$
$$= \sum_{r=1}^{p} \sum_{s=1}^{p} (x_r - \bar{x}_{ri})c^{rs}(x_s - \bar{x}_{si}) \qquad (8.1)$$

where c^{rs} is the element in the rth row and sth column of $\mathbf{C}^{-1}$. The observation $\mathbf{x}$ is allocated to the group for which D_i^2 has the smallest value.

8.3 Canonical discriminant functions

It is sometimes useful to be able to determine functions of the variables $X_1, X_2, \ldots, X_p$ that in some sense separate the m groups as well as is possible. The simplest approach involves taking a linear combination of the X variables

$$Z = a_1 X_1 + a_2 X_2 + \cdots + a_p X_p$$

for this purpose. Groups can be well separated using Z if the mean value changes considerably from group to group, with the values within a group being fairly constant. One way to choose the coefficients $a_1, a_2, \ldots, a_p$ in the index is therefore so as to maximize the F ratio for a one-way analysis of variance. Thus if there are a total of N individuals in all the groups, an analysis of variance on Z values takes the form shown in the following table:

Source of variation	Degrees of freedom	Mean square	F ratio
Between groups	$m - 1$	M_B	M_B/M_W
Within groups	$N - m$	M_W	
	$N - 1$		

Hence a suitable function for separating the groups can be defined as the linear combination for which the F ratio M_B/M_W is as large as possible. This idea was first used by Fisher (1936).

When this approach is used, it turns out that it may be possible to determine several linear combinations for separating groups. In general, the number available is the smaller of p and $m - 1$, say s. They are referred to as *canonical* discriminant functions. The first function

$$Z_1 = a_{11} X_1 + a_{12} X_2 + \cdots + a_{1p} X_p$$

gives the maximum possible F ratio on a one-way analysis of variance for the variation within and between groups. If there is more than one function then the second one

$$Z_2 = a_{21} X_1 + a_{22} X_2 + \cdots + a_{2p} X_p$$

gives the maximum possible F ratio on a one-way analysis of variance subject to the condition that there is no correlation between Z_1 and

Z_2 within groups. Further functions are defined in the same way. Thus the ith canonical discriminant function

$$Z_i = a_{i1}X_1 + a_{i2}X_2 + \cdots + a_{ip}X_p$$

is the linear combination for which the F ratio on an analysis of variance is maximized, subject to Z_i being uncorrelated with $Z_1, Z_2, \ldots$, and Z_{i-1} within groups.

Finding the coefficients of the canonical discriminant functions turns out to be an eigenvalue problem. The within-sample matrix of sums of squares and cross products, $\mathbf{W}$, is calculated using equation (4.12), and $\mathbf{T}$, the total sample matrix of sums of squares and cross products is calculated using equation (4.11). From these, the between-groups matrix

$$\mathbf{B} = \mathbf{T} - \mathbf{W}$$

can be determined. Next, the eigenvalues and eigenvectors of the matrix $\mathbf{W}^{-1}\mathbf{B}$ have to be found. If the eigenvalues are $\lambda_1 > \lambda_2 > \cdots > \lambda_s$ then λ_i is the ratio of the between-group sum of squares to the within-group sum of squares for the ith linear combination, Z_i, while the elements of the corresponding eigenvector, $\mathbf{a}'_i = (a_{i1}, a_{i2}, \ldots, a_{ip})$, are the coefficients of Z_i.

The canonical discriminant functions $Z_1, Z_2, \ldots, Z_s$ are linear combinations of the original variables chosen in such a way that Z_1 reflects group differences as much as possible; Z_2 captures as much as possible of the group differences not displayed by Z_1; Z_3 captures as much as possible of the group differences not displayed by Z_1 and Z_2; etc. The hope is that the first few functions are sufficient to account for almost all of the important group differences. In particular, if only the first one or two functions are needed for this purpose then a simple graphical representation of the relationship between the various groups is possible by plotting the values of these functions for the sample individuals.

8.4 Tests of significance

Several tests of significance are useful in conjunction with a discriminant function analysis. In particular, the T^2 test of equation (4.5)

Table 8.2 Means and standard deviations for the canonical discriminant function Z_1 for five samples of Egyptian skulls

Sample	Mean	Std. dev.
Early predynastic	−0.029	0.097
Late predynastic	−0.043	0.071
12th & 13th dynasties	−0.099	0.075
Ptolemaic	−0.143	0.080
Roman	−0.167	0.095

lue of the discriminant function in this case. The means and
deviations found for the Z_1 values for the five samples are
Table 8.2. It can be seen quite clearly that the mean of Z_1
me lower over time, indicating a trend towards shorter,
skulls with short jaws but relatively large nasal heights.
wever, very much an average change. If the 150 skulls are
to the samples to which they are closest according to the
bis distance function of equation (8.1) then only a fairly
ortion are allocated to the samples that they really belong
8.3). Thus although this discriminant function analysis
uccessful in pinpointing the changes in skull dimensions
it has not produced a satisfactory method for 'ageing'

.3 Results obtained when 150 Egyptian skulls are al-
to the groups for which they have the minimum
nobis distance

	Number allocated to group					
1	2	3	4	5	Total	
12	8	4	4	2	30	
10	8	5	4	3	30	
4	4	15	2	5	30	
3	3	7	5	12	30	
2	4	4	9	11	30	

can be used to test for a significant difference between the mean values for any pair of groups, while the likelihood ratio test of equation (4.10) can be used to test for overall differences between the means of the m groups.

In addition, a test is sometimes proposed for testing whether the mean of the discriminant function Z_j differs significantly from group to group. This is based on the statistic

$$\phi_j^2 = \left\{ \sum_{i=1}^{m} n_i - 1 - \tfrac{1}{2}(p + m) \right\} \log_e(1 + \lambda_j),$$

which is tested against the chi-squared distribution with $p + m - 2j$ degrees of freedom. A significantly large value is taken as evidence that group differences exist. Alternatively, the sum $\lambda_j^2 + \lambda_{j+1}^2 + \cdots + \lambda_s^2$ is sometimes proposed for testing for group differences related to discriminant functions Z_j to Z_s. This is tested against the chi-squared distribution with the sum of the degrees of freedom associated with the component terms.

Unfortunately, these tests are highly suspect because the jth discriminant function in the population may not appear as the jth discriminant function in the sample as a result of sampling errors. For example, the estimated first discriminant function (corresponding to the largest eigenvalue for the sample matrix $\mathbf{W}^{-1}\mathbf{B}$) may in reality correspond to the second discriminant function for the population being sampled. Simulations indicate that this can upset the chi-squared tests using λ_j statistics and sums of these statistics very seriously. Therefore it seems that the tests should not be relied upon the decide how many of the obtained discriminant functions represent real group differences. See Harris (1985) for an extended discussion of the difficulties surrounding these tests, and alternative ways for examining the nature of group differences.

One useful type of test that is valid, at least for large samples, involves calculating the Mahalanobis distance from each of the observations to the mean vector for the group containing the observation as discussed in section 5.3. These distances should follow approximately chi-squared distributions with p degrees of freedom. Hence if an observation is very significantly far from the centre of its group in comparison from the chi-squared distribution then this brings into question whether the observation really came from the group.

8.5 Assumptions

The methods discussed so far in this chapter are based on two assumptions. First, for all the methods, the within-group covariance matrix should be the same for all groups. Second, for tests of significance, the data should be normally distributed within groups.

In general it seems that multivariate analyses that assume normality may be upset quite badly if this assumption is not correct. This contrasts with the situation with univariate analyses such as regression and analysis of variance, which are generally quite robust to this assumption. However, a failure of one or both assumptions does not necessarily mean that a discriminant function analysis as described above is a waste of time. For example, it may well turn out that excellent discrimination is possible on data from non-normal distributions, although it may not then be simple to establish the statistical significance of the group differences. Furthermore, discrimination methods that do not require the assumptions of normality and equality of population covariance matrices are available (see below).

Example 8.1 Comparison of samples of Egyptian skulls

This example concerns the comparison of the values for four measurements on male Egyptian skulls for five samples ranging in age from the early predynastic period (*circa* 4000 BC) to the Roman period (*circa* AD 150). The data are shown in Table 1.2 and it has already been established that the mean values differ significantly from sample to sample (Example 4.3), with the differences tending to increase with the time difference between samples (Example 5.3).

The within-sample matrix of sums of squares and cross products is calculated using equation (4.12) as

$$\mathbf{W} = \begin{bmatrix} 3061.67 & 5.33 & 11.47 & 291.30 \\ 5.33 & 3405.27 & 754.00 & 412.53 \\ 11.47 & 754.00 & 3505.97 & 164.33 \\ 291.30 & 412.53 & 164.33 & 1472.13 \end{bmatrix},$$

while the corresponding total sample matrix is calculated using equation (4.11) as

$$\mathbf{T} = \begin{bmatrix} 3563.89 & -222.81 & -61 \\ -222.81 & 3635.17 & 104 \\ -615.16 & 1046.28 & 43 \\ 426.73 & 346.47 & - \end{bmatrix}$$

The between-sample matrix is therefore

$$\mathbf{B} = \mathbf{T} - \mathbf{W} = \begin{bmatrix} 502.83 & -228.15 \\ -228.15 & 229.9 \\ -626.63 & 292.2 \\ 135.43 & -66. \end{bmatrix}$$

The eigenvalues of $\mathbf{W}^{-1}\mathbf{B}$ are four $\lambda_3 = 0.015$ and $\lambda_4 = 0.002$. The corres functions are

$$Z_1 = -0.0107X_1 + 0.0040X_2 +$$

$$Z_2 = 0.0031X_1 + 0.0168X_2$$

$$Z_3 = -0.0068X_1 + 0.0010X_2$$

and

$$Z_4 = 0.0126X_1 - 0.0001X$$

Because λ_1 is much larger than most of the sample differences

The X variables in equatic Table 1.2 without standardiz from which it can be seen t skulls that are tall but narrc height.

The Z_1 values for individ For example, the first indi has $X_1 = 131$ mm, $X_2 = 1$ Therefore

$$Z_1 = -0.0107 \times$$

$$- 0.0068 \times$$

is the va standard shown in has beco broader This is, h allocated Mahaland small prop to (Table has been s over time, skulls.

Table 8 located Mahala

Source group

1
2
3
4
5

Example 8.2 Discriminating between groups of European countries

The data shown in Table 1.5 on employment percentages in nine groups in 26 European countries have already been examined by principal components analysis and by factor analysis (Examples 6.2 and 7.1). Here they will be considered from the point of view of the extent to which it is possible to discriminate between groups of countries on the basis of employment patterns. In particular, three natural groups existed in 1979 when the data were collected. These were: (1) the European Economic Community (EEC) countries at the time of Belgium, Denmark, France, West Germany, Ireland, Italy, Luxembourg, the Netherlands and the United Kingdom; (2) the other western European countries of Austria, Finland, Greece, Norway, Portugal, Spain, Sweden, Switzerland and Turkey; and (3) the eastern European communist countries of Bulgaria, Czechoslovakia, East Germany, Hungary, Poland, Romania, USSR and Yugoslavia. These three groups can be used as a basis for a discriminant function analysis.

The percentages in the nine industry groups add to 100% for each of the 26 countries. This means that any one of the nine percentage variables can be expressed as 100 minus the remaining variables. It is therefore necessary to omit one of the variables from the analysis in order to calculate Mahalanobis distances and canonical discriminant functions. The last variable, the percentage employed in transport and communications, was omitted for the analysis that will now be described.

The number of canonical variables is two in this example, this being the minimum of the number of variables ($p = 8$) and the number of groups minus one ($m - 1 = 2$). These canonical variables are

$$Z_1 = 0.73\,\text{AGR} + 0.62\,\text{MIN} + 0.63\,\text{MAN} - 0.16\,\text{PS}$$
$$+ 0.80\,\text{CON} + 1.24\,\text{SER} + 0.72\,\text{FIN} + 0.82\,\text{SPS},$$

and

$$Z_2 = 0.84\,\text{AGR} + 2.46\,\text{MIN} + 0.78\,\text{MAN} + 1.18\,\text{PS}$$
$$+ 1.17\,\text{CON} + 0.83\,\text{SER} + 0.84\,\text{FIN} + 1.05\,\text{SPS},$$

the corresponding eigenvalues of $\mathbf{W}^{-1}\mathbf{B}$ being $\lambda_1 = 7.531$ and $\lambda_2 = 1.046$.

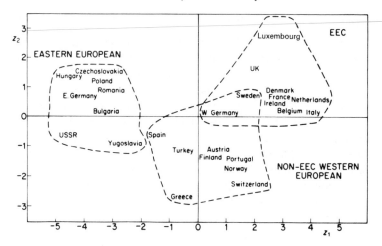

Figure 8.1 Plot of 26 European countries against their values for two canonical discriminant functions.

From the coefficients in the equation for Z_1 it can be seen that this variable will tend to be large when there are high percentages employed in everything except PS (power supplies). There is a particularly high coefficient for SER (service industries). For Z_2, on the other hand, all the coefficients are positive, with that for MIN (mining) being particularly high.

A plot of the countries against their values for Z_1 and Z_2 is shown in Fig. 8.1. The eastern European communist countries appear on the left-hand side, the non-EEC western European countries in the centre, and the EEC countries on the right-hand side of the figure. It can be clearly seen how most separation occurs with the horizontal values of Z_1. As far as values of Z_2 are concerned, it appears that the non-EEC western European countries tend to have lower values than the other two groups. Overall, the degree of separation of the three groups is good. The only 'odd' cases are West Germany, which appears to be more like a non-EEC western European country than an EEC country, and Sweden, which appears to be more like an EEC country than a non-EEC western European country.

The discriminant function analysis has been rather successful in this example. It is possible to separate the three groups of countries on the basis of their employment patterns. Furthermore, the separation using the two canonical discriminant functions is much clearer than the separation shown in Fig. 6.2 for the first two principal components.

8.6 Allowing for prior probabilities of group membership

Computer programs often allow many options for a discriminant function analysis. One situation is that the probability of membership is inherently different for different groups. For example, if there are two groups it might be that it is known that most individuals fall into group 1 while very few fall into group 2. In that case if an individual is to be allocated to a group it makes sense to bias the allocation procedure in favour of group 1. Thus the process of allocating an individual to the group to which it has the smallest Mahalanobis distance should be modified. To allow for this some computer programs permit prior probabilities of group membership to be taken into account in the analysis.

8.7 Stepwise discriminant function analysis

Another possible modification of the basic analysis involves carrying it out in a stepwise manner. In this case variables are added to the discriminant functions one by one until it is found that adding extra variables does not give significantly better discrimination. There are many different criteria that can be used for deciding on which variables to include in the analysis and which to miss out.

A problem with stepwise discriminant function analysis is the bias that the procedure introduces into significance tests. Given enough variables it is almost certain that some combination of them will produce significant discriminant functions by chance alone. If a stepwise analysis is carried out then it is advisable to check its validity by rerunning it several times with a random allocation of individuals to groups to see how significant are the results obtained. For example, with the Egyptian skull data the 150 skulls could be allocated completely at random to five groups of 30, the allocation being made a number of times, and a discriminant function analysis run on each random set of data. Some idea could then be gained of the probability of getting significant results through chance alone.

It should be stressed that this type of randomization to verify a discriminant function analysis is unnecessary for a standard non-stepwise analysis providing there is no reason to suspect the assumptions behind the analysis. It could, however, be informative in cases where the data are clearly not normally distributed within groups or where the within-group covariance matrix is not the same for each group.

8.8 Jackknife classification of individuals

A moment's reflection will suggest that an allocation matrix such as that shown in Table 8.3 must tend to have a bias in favour of allocating individuals to the group that they really come from. After all, the group means are determined from the observations in that group. It is not surprising to find that an observation is closest to the centre of the group where the observation helped to determine that centre.

To overcome this bias, some computer programs carry out what is called a 'jackknife classification' of observations. This involves allocating each individual to its closest group without using that individual to help determine a group centre. In this way any bias in the allocation is avoided. In practice there is often not a great deal of difference between the straightforward classification and the jackknife classification. The jackknife classification usually gives a slightly smaller number of correct allocations.

8.9 Assigning of ungrouped individuals to groups

Some computer programs allow the input of data values for a number of individuals for which the true group is not known. It is then possible to assign these individuals to the group that they are closest to, in the Mahalanobis distance sense, on the assumption that they have to come from one of the m groups that are sampled. Obviously in these cases it will not be known whether the assignment is correct. However, the errors in the allocation of individuals from known groups is an indication of how accurate the assignment process is likely to be. For example, the results shown in Table 8.3 indicate that allocating Egyptian skulls to different time periods using skull dimensions is liable to result in many errors.

8.10 Logistic regression

A rather different approach to discrimination between two groups involves making use of logistic regression. In order to explain how this is done, the more usual use of logistic regression will first be briefly explained because this not commonly covered in introductory statistics courses.

The general framework for logistic regression is that there are m groups to be compared, with group i consisting of n_i items, of which Y_i exhibit a positive response (a 'success') and $n_i - Y_i$ exhibit a negative response (a 'failure'). The assumption is then made that the probability of a 'success' for an item in group i is given by

$$\pi_i = \frac{\exp(\beta_0 + \beta_1 x_{i1} + \beta_2 x_{i2} + \cdots + \beta_p x_{ip})}{1 + \exp(\beta_0 + \beta_1 x_{i1} + \beta_2 x_{i2} + \cdots + \beta_p x_{ip})}, \tag{8.3}$$

where x_{ij} is the value of some variable X_j that is the same for all items in the group. In this way the variables X_1 to X_p are allowed to influence the probability of a 'success', which is assumed to be the same for all items in the group, irrespective of the 'successes' or 'failures' of the other items in that or any other group. Similarly, the probability of a 'failure' is $1 - \pi_i$ for all items in the ith group. It is permissible for some or all of the group to contain just one item. Indeed, some computer programs only allow for this to be the case.

There need be no concern about arbitrarily choosing what to call a 'success' and what to call a 'failure'. It is easy to show that reversing these designations in the data simply results in all the β values and their estimates changing sign, and consequently π_i changing to $1 - \pi_i$.

The function that is used to relate the probability of a success to the X variables is called a logistic function. Unlike the standard multiple regression function, the logistic function forces estimated probabilities to lie within the range zero to one. It is for this reason that logistic regression is more sensible than linear regression as a means of modelling probabilities.

There are numerous computer programs available for fitting equation (8.3) to data, i.e. for estimating the values of β_0 to β_p. Commonly they are based on the principle of maximum likelihood, which means that the equations for the estimation of the β values do not have an explicit solution. As a result, the calculations involve an iterative process of improving initial approximations for the estimates until no further changes can be made. The output commonly includes the estimates of the β values and their standard errors, a chi-squared statistic that indicates the extent to which the model fits the data, and a chi-squared statistic which indicates the extent to which the model is an improvement over what is obtained by assuming that the probability of a 'success' is unrelated to the X variables.

In the context of discrimination with two samples there are three different types of situation that have to be considered:

1. The data consist of a single random sample taken from a population of items that is itself divided into two parts. The application of logistic regression is then straightforward and the fitted equation (8.3) can be used to give an estimate of the probability of an item being in one part of the population (i.e. is a 'success') as a function of the values that the item possesses for variables X_1 to X_p. In addition, the distribution of 'success' probabilities for the sampled items is an estimate of the distribution of these probabilities for the full population.
2. Separate sampling is used where a random sample of size n_1 is taken from the population of items of one type (the 'successes'), and an independent random sample of size n_2 is taken from the population of items of the second type (the 'failures'). Logistic regression can still be used. However, the estimated probability of a 'success' obtained from the estimated function must be interpreted in terms of the sampling scheme and the sample sizes used.
3. Groups of items are chosen to have particular values for the variables X_1 to X_p, such that these variable values change from group to group. The number of 'successes' in each group is then observed. In this case the estimated logistic regression equation gives the probability of a 'success' for an item, conditional on the values that the item possesses for X_1 to X_p. The estimated function is therefore the same as for situation (1), but the sample distribution of probabilities of a 'success' is in no way an estimate of the distribution that would be found in the combined population of items that are 'successes' or 'failures'.

The following examples illustrate the differences between situations (1) and (2), which are the ones that most commonly occur. Situation (3) is eally just a standard logistic regression situation, and will not be considered further here.

Example 8.3 Storm survival of female sparrows (reconsidered)

The data in Table 1.1 consist of values for five morphological variables for 49 female sparrows taken in a moribund condition to Hermon Bumpus's laboratory at Brown University, Rhode Island, after a severe storm in 1898. The first 21 birds recovered and the remaining 28 died, and there is some interest in knowing whether it is possible to discriminate between these two groups on the basis of the five measurements. It has already been shown that there are no significant differences between the mean values of the variables for survivors and non-survivors (Example 4.1), although the non-survivors may have been more variable (Example 4.2). A principal components analysis has also confirmed the test results (Example 6.1).

This is a situation of type (1) if the assumption is made that the sampled birds were randomly selected from the population of female sparrows in some area close to Bumpus's laboratory. Actually, the assumption of random sampling is questionable because it is not clear exactly how the birds were collected. Nevertheless, the assumption will be made for this example.

The logistic regression option in the program SOLO (BMDP, 1989) was used to fit the model

$$\pi_i = \frac{\exp(\beta_0 + \beta_1 x_{i1} + \beta_2 x_{i2} + \cdots + \beta_p x_{i5})}{1 + \exp(\beta_0 + \beta_1 x_{i1} + \beta_2 x_{i2} + \cdots + \beta_p x_{i5})},$$

where the variables are X_1 = total length, X_2 = alar extent, X_3 = length of beak and head, X_4 = length of humerus and X_5 = length of sternum, all in mm, and π_i denotes the probability of the ith bird recovering from the storm.

A chi-squared test for whether the variables account significantly for the difference between survivors and non-survivors gives the value 2.85 with five degrees of freedom, which is not at all significantly large when compared with chi-squared tables. There is therefore no evidence from this analysis that the survival status was related to the morphological variables. Estimated values for β_0 to β_5 are shown in Table 8.4, together with estimated standard errors and chi-squared statistics for testing whether the individual estimates differ significantly from zero. Again, there is no evidence of any significant effects.

The effect of adding X_1^2 to X_5^2 to the model was also investigated. This did not introduce any significant results. Adding the 10 product

Table 8.4 Estimates of the constant term and the coefficients of X variables when a logistic regression model is fitted to data on the survival of 49 female sparrows. The chi-squared value is (estimated β value/standard error)^2. The P-value is the probability of a value this large from the chi-squared distribution with one degree of freedom. A small P-value (say less than 0.05) provides evidence that the true value of the β value concerned is not zero

Variable	β estimate	Standard error	Chi-squared value	P-value
β_0	13.582	15.865	–	–
β_1	−0.163	0.140	1.36	0.244
β_2	−0.028	0.106	0.07	0.794
β_3	−0.084	0.629	0.02	0.894
β_4	1.062	1.023	1.08	0.299
β_5	0.072	0.417	0.03	0.864

terms $X_1 \times X_2$, $X_1 \times X_3, \ldots, X_4 \times X_5$ as well as the squared terms was also investigated. In this case SOLO failed to fit the logistic function, probably because there were then 21 β values to be estimated using only 49 data points.

In summary, logistic regression gives no indication at all that the survival of the female sparrows was related to the measured variables.

Example 8.4 Comparison of two samples of Egyptian skulls

As an example of separate sampling, where the sample size in the two groups being compared is not necessarily related in any way to the respective population sizes, consider the comparison between the first and last samples of Egyptian skulls for which data are provided in Table 1.2. The first sample consists of 30 male skulls from burials in the area of Thebes during the early predynastic period (*circa* 4000 BC) in Egypt, and the last sample consists of 30 male skulls from burials in the same area during the Roman period (*circa* AD 150). For each skull, measurements are available for $X_1 =$ maximum breadth, $X_2 =$ basibregmatic height, $X_3 =$ basialveolar length and $X_4 =$ nasal height, all in mm (Fig. 1.1). For the purpose of this example it will be assumed that the two samples were effectively randomly chosen from their respective populations, although there is no way of knowing how realistic this is.

Obviously the equal sample sizes in no way indicate that the population sizes in the two periods were equal. They are in fact completely arbitrary because many more skulls have been measured from both periods and an unknown number of skulls have either not survived intact or not been found. Therefore, if the two samples are lumped together and treated as a 'sample' of size 60 for the estimation of a logistic regression equation then it is clear that the estimated 'probability' of a skull with certain dimensions being from the early predynastic period may not really be estimating the true probability at all.

In fact, it is difficult to define precisely what is meant by the 'true probability' in this example because the population is not at all clear. A working definition is that the probability of a skull with specified dimensions being from the predynastic period is equal to the proportion of all skulls with the given dimensions that are from the predynastic period in a hypothetical population of all male skulls from either the predynastic or the Roman period that might have been recovered by archaeologists in the Thebes region.

It can be shown (Seber, 1984, p. 312) that if a logistic regression is carried out on a lumped sample to estimate the equation (8.3), then the modified equation

$$\tau_i = \frac{\exp(\beta_0 - \log_e\{(n_1 P_2)/(n_2 P_1)\} + \beta_1 x_{i1} + \beta_2 x_{i2} + \cdots + \beta_p x_{ip})}{1 + \exp(\beta_0 - \log_e\{(n_1 P_2)/(n_2 P_1)\} + \beta_1 x_{i1} + \beta_2 x_{i2} + \cdots + \beta_p x_{ip})},$$

(8.4)

is what really gives the probability that an item with the specified X values is a 'success'. Here equation (8.4) differs from equation (8.3) because of the term $\log_e\{(n_1 P_2)/(n_2 P_1)\}$ in the numerator and the denominator, where P_1 is the proportion of items in the full population of 'successes' and 'failures' that are 'successes', and $P_2 = 1 - P_1$ is the proportion of the population that are 'failures'. This then means that in order to estimate the probability of an item with the specified X values being a 'success' requires that P_1 and P_2 are either known or can somehow be estimated separately from the sample data in order to adjust the estimated logistic regression equation for the fact that the sample sizes n_1 and n_2 are not proportional to the population frequencies of 'successes' and 'failures'. In the example being considered this requires that estimates of the relative frequencies of predynastic and Roman skulls in the Thebes area must be known in order to be able to estimate the probability that a given skull is predynastic, given the values that it possesses for the variables X_1 to X_4.

The logistic regression option of SOLO (BMDP, 1989) was used with the lumped data from the 60 predynastic and Roman skulls, with a predynastic skull being treated as a 'success'. The resulting chi-squared test statistic output from SOLO as a measure of the extent to which 'success' is related to the X variables is 27.13 with four degrees of freedom. This is significantly large at the 0.1% level, giving very strong evidence of a relationship. The estimates of the constant term and the coefficients of the X variables are shown in Table 8.5. It can be seen that the estimate of β_1 is significantly different from zero at about the 1% level and β_3 is significantly different from zero at the 2% level. Hence X_1 and X_3 appear to be the important variables for discriminating between the two types of skull.

The fitted function can be used to discriminate between the two groups of skulls by assigning values for P_1 and $P_2 = 1 - P_1$ in equation (8.4). As already noted, it is desirable that these should correspond to the population proportions of predynastic and Roman skulls. However, this is not possible because these proportions are not known. In practice, therefore, arbitrary values must be assigned. For example, suppose that P_1 and P_2 are both set equal to $\frac{1}{2}$. Then $\log_e\{(n_1 P_2)/(n_2 P_1)\} = \log_e(1) = 0$ because $n_1 = n_2$, and equations (8.3) and (8.4) become identical. The logistic function therefore estimates the probability of a skull being predynastic in a population with equal frequencies of predynastic and Roman skulls.

The extent to which the logistic equation is effective for discrimination is indicated in Fig. 8.2, which shows the estimated values

Table 8.5 Estimates β_0 and the coefficients of X variables when a logistic regression model is fitted to data on 30 predynastic and 30 Roman period male Egyptian skulls. See Table 8.4 for an explanation of the other columns

Variable	β estimate	Standard error	Chi-squared value	P-value
β_0	−6.723	13.081	–	–
β_1	−0.202	0.075	7.13	0.008
β_2	0.129	0.079	2.66	0.103
β_3	0.177	0.073	5.84	0.016
β_4	−0.008	0.104	0.01	0.939

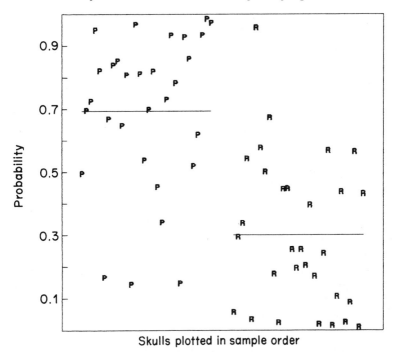

Figure 8.2 Values from a fitted logistic function plotted for 30 predynastic (P) and 30 Roman (R) skulls. The horizontal lines indicate the average probabilities for the two groups.

of π_i for the 60 sample skulls. There is a distinct difference in the distributions of values for the two samples, with the mean for predynastic skulls being about 0.7 and the mean for Roman skulls being about 0.3. However, there is also a considerable overlap between the distributions. As a result, if the sample skulls are classified as being predynastic when the logistic equation gives a value greater than 0.5 or as Roman when the equation gives a value of less than 0.5, then six predynastic skulls are misclassified as being Roman, and seven Roman skulls are misclassified as being predynastic.

8.11 Computational methods and computer programs

Major statistical packages generally have a discriminant function option that applies the methods described in sections 8.2–8.4, based on the assumption of normally distributed data. However, the details of the order of calculations, the way the output is given, and the terminology vary considerably. Therefore, manuals have to be studied carefully to determined precisely what is done by any individual program.

Logistic regression is also fairly widely available. In some programs such as SOLO (BMDP, 1989) there is the restriction that all items are assumed to have different values for X variables. However, it is more common for groups of items with common X values to be permitted.

8.12 Further reading

The asumption that samples are from multivariate distributions with the same covariance matrix that is required for the use of the methods described in sections 8.2–8.4 can be relaxed. If the samples being compared are assumed to come from multivariate normal distributions with different covariance matrices then a method called quadratic discriminant function analysis can be applied. This option is available in the computer packages MINITAB (1989) and SAS (1985). See Seber (1984, p. 297) for more information about this method and a discussion of its performance relative to the more standard linear discriminant function analysis.

Discrimination using logistic regression has been described in section 8.10 in terms of the comparison of two groups. More detailed treatments of this method are provided by Hosmer and Lemeshow (1989) and Collett (1991). The method can be generalized for discrimination between more than two groups if necessary. The BMDP computer package (Dixon, 1990) has this as an option.

Exercises

1. Consider the data in Table 4.4 for nine mandible measurements on samples from five different canine groups. Carry out a discriminant function analysis to see how well it is possible to separate the groups using the measurements.
2. Still considering the data in Table 4.4, investigate each canine group separately to see whether logistic regression shows a significant difference between males and females for the measurements. Note that in view of the small sample sizes available for each group it is unreasonable to expect to fit a logistic function

involving all nine variables with good estimates of parameters. Therefore consideration should be given to fitting functions using only a subset of the variables.

References

BMDP (1989) *SOLO User's Guide, Version 3.1.* BMDP Statistical Software, 1440 Sepulveda Boulevard, Los Angeles, California 90025.

Collett, D. (1991) *Modelling Binary Data.* Chapman and Hall, London.

Dixon, W.J. (ed.) (1990) *BMDP Statistical Software Manual.* University of California Press, Los Angeles.

Fisher, R.A. (1936) The utilization of multiple measurements in taxonomic problems. *Annals of Eugenics* 7, 179–88.

Harris, R.J. (1985) *A Primer on Multivariate Statistics,* 2nd edn. Academic Press, Orlando.

Hosmer, D.W. and Lemeshow, S. (1989) *Applied Logistic Regression.* Wiley, New York.

MINITAB (1989) *MINITAB Reference Manual, Release 7.* Minitab Inc., 3081 Enterprise Drive, State College, Pennsylvania 16801.

SAS (1985) *SAS User's Guide: Statistics.* SAS Institute, Cary, North Carolina 27511.

Seber, G.A.F. (1984) *Multivariate Observations.* Wiley, New York.

CHAPTER NINE

Cluster analysis

9.1 Uses of cluster analysis

The problem that cluster analysis is designed to solve is the following one: given a sample of n objects, each of which has a score on p variables, devise a scheme for grouping the objects into classes so that 'similar' ones are in the same class. The method must be completely numerical and the number of classes is not known. This problem is clearly more difficult than the problem for a discriminant function analysis because with discriminant function analysis the groups are known to begin with.

There are many reasons why cluster analysis may be worth while. Firstly, it might be a question of finding the 'true' groups. For example, in psychiatry there has been a great deal of disagreement over the classification of depressed patients, and cluster analysis has been used to define 'objective' groups. Secondly, cluster analysis may be useful for data reduction. For example, a large number of cities can potentially be used as test markets for a new product but it is only feasible to use a few. If cities can be placed into a small number of groups of similar cities then one member from each group can be used for the test market. On the other hand, if cluster analysis generates unexpected groupings then this might in itself suggest relationships to be investigated.

9.2 Types of cluster analysis

Many algorithms have been proposed for cluster analysis. Here attention will be restricted to those following two particular approaches. Firstly, there are hierarchic techniques which produce a *dendrogram* such as the ones shown in Fig. 9.1. These methods start with the calculation of the distances of each individual to all other individuals. Groups are then formed by a process of agglomeration

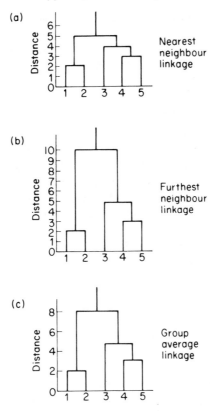

Figure 9.1 Examples of dendograms from cluster analyses of five objects.

or division. With agglomeration all objects start by being alone in groups of one. Close groups are then gradually merged until finally all individuals are in a single group. With division all objects start in a single group. This is then split into two groups, the two groups are then split, and so on until all objects are in groups of their own.

The second approach to cluster analysis involves partitioning, with objects being allowed to move in and out of groups at different stages of the analysis. To begin with, some arbitrary group centres are chosen and individuals are allocated to the nearest one. New centres are then calculated where these are at the centres of the individuals in groups. An individual is then moved to a new group if it is closer to that group's centre than it is to the centre of its present group. Groups 'close' together are merged; spread-out groups are split, etc. The process continues iteratively until stability is achieved with a predetermined number of groups. Usually a range of values is tried for the final number of groups.

9.3 Hierarchic methods

Agglomerative hierarchic methods start with a matrix of 'distances' between individuals. All individuals begin alone in groups of size one and groups that are 'close' together are merged. (Measures of 'distance' will be discussed later.) There are various ways to define 'close'. The simplest is in terms of *nearest neighbours*. For example, suppose there is the following distance matrix for five objects:

	1	2	3	4	5
1	–				
2	2	–			
3	6	5	–		
4	10	9	4	–	
5	9	8	5	3	–

The calculations are then as shown in the following table. Groups are merged at a given level of distance if one of the individuals in one group is that distance or closer to at least one individual in the second group.

Distance	Groups
0	1, 2, 3, 4, 5
2	(1, 2), 3, 4, 5
3	(1, 2), 3, (4, 5)
4	(1, 2), (3, 4, 5)
5	(1, 2, 3, 4, 5)

At a distance of 0 all five objects are on their own. The distance matrix shows that the smallest distance between two objects is 2,

which is between the first and second objects. Hence at a distance
level of 2 there are four groups (1, 2), (3), (4) and (5). The next smallest
distance between objects is 3, which is between objects 4 and 5.
Hence at a distance of 3 there are three groups (1, 2), (3) and (4, 5).
The next smallest distance is 4, which is between objects 3 and 4.
Hence at this level of distance there are two groups (1, 2) and (3, 4, 5).
Finally, the next smallest distance is 5, which is between objects 2
and 3 and between objects 3 and 5. At this level the two groups
merge into the single group (1, 2, 3, 4, 5) and the analysis is complete.
The dendrogram shown in Fig. 9.1(a) illustrates how agglomeration
takes place.

With *furthest neighbour* linkage two groups merge only if the
most distant members of the two groups are close enough. With the
example data this works as follows:

Distance	Groups
0	1, 2, 3, 4, 5
2	(1, 2), 3, 4, 5
3	(1, 2), 3, (4, 5)
5	(1, 2), (3, 4, 5)
10	(1, 2, 3, 4, 5)

Object 3 does not join with objects 4 and 5 until distance level 5
because this is the distance to object 3 from the furthest away of
objects 4 and 5. The furthest-neighbour dendrogram is shown in
Fig. 9.1(b).

With *group average* linkage two groups merge if the average
distance between them is small enough. With the example data this
gives the following result:

Distance	Groups
0	1, 2, 3, 4, 5
2	(1, 2), 3, 4, 5
3	(1, 2), 3, (4, 5)
4.5	(1, 2), (3, 4, 5)
7.8	(1, 2, 3, 4, 5)

For instance, groups (1, 2) and (3, 4, 5) merge at distance level 7.8 as
this is the average distance from objects 1 and 2 to objects 3, 4 and
5, the actual distances being:

1–3	6
1–4	10
1–5	9
2–3	5
2–4	9
2–5	8

Mean $= 7.8$

The dendrogram in this case is shown in Fig. 9.1(c).

Divisive hierarchic methods have been used less often than agglomerative ones. The objects are all put into one group initially, and
then this is split into two groups by separating off the object that
is furthest on average from the other objects. Individuals from the
main group are then moved to the new group if they are closer to
it than they are to the main group. Further subdivisions occur as
the distance that is allowed between individuals in the same group
is reduced. Eventually all objects are in groups of size one.

9.4 Problems of cluster analysis

It has already been mentioned that there are many algorithms
for cluster analysis. However, there is no generally accepted 'best'
method. Unfortunately, different algorithms do not necessarily produce the same results on a given set of data and there is usually
rather a large subjective component in the assessment of the results
from any particular method.

A fair test of any algorithm is to take a set of data with a known
group structure and see whether the algorithm is able to reproduce
this structure. It seems to be the case that this test only works in cases
where the groups are very distinct. When there is a considerable
overlap between the initial groups, a cluster analysis may produce
a solution that is quite different from the true situation.

In some cases difficulties will arise because of the shape of clusters.
For example, suppose that there are two variables X_1 and X_2 and
individuals are plotted according to their values for these. Some

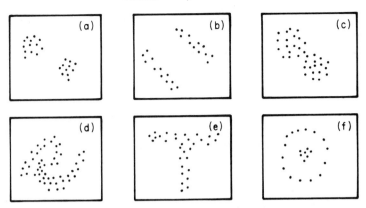

Figure 9.2 Some possible patterns of points with two clusters.

possible patterns of points are illustrated in Fig. 9.2. Case (a) is likely to be found by any reasonable algorithm, as is case (b). In case (c) some algorithms might well fail to detect two clusters because of the intermediate points. Most algorithms would have trouble handling cases like (d), (e) and (f).

Of course, clusters can only be based on the variables that are given in the data. Therefore they must be relevant to the classification wanted. To classify depressed patients there is presumably not much point in measuring height, weight, or length of arms. A problem here is that the clusters obtained may be rather sensitive to the particular choice of variables that is made. A different choice of variables, apparently equally reasonable, may give different clusters.

9.5 Measures of distance

The data for a cluster analysis usually consists of the values of p variables $X_1, X_2, \ldots, X_p$ for n objects. For hierarchic algorithms these values are then used to produce an array of distances between the individuals. Measures of distance have already been discussed in Chapter 5. Here it suffices to say that the Euclidean distance function

$$d_{ij} = \sqrt{\left\{ \sum_{k=1}^{p} (x_{ik} - x_{jk})^2 \right\}} \tag{9.1}$$

is most frequently used for quantitative variables, where x_{ik} is the value of variable X_k for individual i and x_{jk} is the value of the same variable for individual j. The geometrical interpretation of the distance d_{ij} from individual i to individual j is illustrated in Figs 5.1 and 5.2 for the cases of two and three variables.

Usually variables are standardized in some way before distances are calculated, so that all p variables are equally important in determining these distances. This can be done by coding so that the means are all zero and the variances are all one. Alternatively, each variable can be coded to have a minimum of zero and a maximum of one. Unfortunately, standardization has the effect of minimizing group differences because if groups are separated well by X_i then the variance of X_i will be large, and indeed it should be large. It would be best to be able to make the variances equal to one within clusters but this is obviously not possible as the whole point of the analysis is to find the clusters.

9.6 Principal components analysis with cluster analysis

Some cluster analysis algorithms begin by doing a principal components analysis to reduce a large number of original variables down to a smaller number of principal components. This can drastically reduce the computing time for the cluster analysis. However, it is known that the results of a cluster analysis can be rather different with and without the initial principal components analysis. Consequently an initial principal components analysis is probably best avoided.

On the other hand, when the first two principal components account for a high percentage of variation in the data a plot of individuals against these two components is certainly a useful way for looking for clusters. For example, Fig. 6.2 shows European countries plotted in this way for principal components based on employment percentages. The countries do seem to group in a meaningful way.

Example 9.1 Clustering of European countries

The data just mentioned on the percentages of people employed in nine industry groups in different countries of Europe (Table 1.5) can be used for a first example of cluster analysis. The analysis should show which countries have similar employment patterns and which countries are different in this respect. It may be recalled from Example

8.2 that a sensible grouping into EEC, non-EEC western European countries and eastern European countries existed in 1979, when the data were collected.

An analysis was carried out using the computer program MVSP (Kovach, 1993). The first step in the analysis involved standardizing the nine variables so that each one had a mean of zero and a standard deviation of one. For example, variable 1 is AGR, the percentage employed in agriculture. For the 26 countries being considered this variable has a mean of 19.13 and a standard deviation of 15.55. The data value for Belgium for AGR is 3.3 which standardizes to $(3.3-19.13)/15.55 = -1.02$; the data value for Denmark is 9.2 which standardizes to -0.64; and so on. The standardized values are shown in Table 9.1.

The next step in the analysis involved calculating the Euclidean distances between all pairs of countries. This was done using equation (9.1) on the standardized data values. Finally, a dendrogram was formed by the agglomerative, nearest neighbour, hierarchic process described above.

The dendrogram is shown in Fig. 9.3. It can be seen that the two closest countries were Sweden and Denmark. These are distance 1.135 apart. The next closest pair of countries are Belgium and France, which are 1.479 apart. Then come Poland and Bulgaria, which are 1.537 apart. Amalgamation ended with Turkey joining the other countries at a distance of 5.019.

Having obtained the dendrogram, we are free to decide how many clusters to take. For example, if six clusters are to be considered then these are found at an amalgamation distance of 2.459. The first cluster is the western nations of Belgium, France, Netherlands, Sweden, Denmark, West Germany, Finland, UK, Austria, Ireland, Switzerland, Norway, Greece, Portugal and Italy. The second cluster is Luxembourg on its own. Then there are the former communist countries of USSR, Hungary, Czechoslovakia, East Germany, Romania, Poland and Bulgaria. The last three clusters are Spain, Yugoslavia and Turkey, each on their own. These clusters do, perhaps, make a certain amount of sense. From the standardized scores shown in Table 9.1 it can be seen that Luxembourg is unusual because of the large numbers in mining. Spain is unusual because of the large numbers in construction. Yugoslavia is unusual because of the large numbers in agriculture and finance and low numbers in construction, social and personal services, and transport and

Table 9.1 Standardized data on the percentages employed in nine industry groups in Europe, as derived from Table 1.5

Country	AGR	MIN	MAN	PS	CON	SER	FIN	SPS	TC
Belgium	-1.02	-0.36	0.09	-0.02	0.02	1.34	0.78	0.96	0.47
Denmark	-0.64	-1.19	-0.74	-0.81	0.08	0.36	0.89	1.78	0.40
France	-0.54	-0.47	0.07	-0.02	0.45	0.84	0.71	0.38	-0.61
West Germany	-0.80	0.05	1.25	-0.02	-0.53	0.32	0.36	0.33	-0.32
Ireland	0.26	-0.26	-0.90	1.04	-0.40	0.84	-0.43	0.11	-0.32
Italy	-0.21	-0.67	0.08	-1.08	1.12	1.12	-0.85	0.01	-0.61
Luxembourg	-0.73	1.90	0.54	-0.29	0.63	1.21	0.21	-0.12	-0.25
Netherlands	-0.82	-1.19	-0.64	0.24	1.05	1.10	1.00	1.24	0.18
UK	-1.06	0.15	0.46	1.31	-0.77	0.86	0.61	1.21	-0.10
Austria	-0.41	-0.16	0.46	1.31	0.51	0.84	0.32	-0.47	0.33
Finland	-0.39	-0.88	-0.16	1.04	-0.46	0.38	0.53	0.63	0.76
Greece	1.43	-0.67	-1.34	-0.82	-0.04	-0.32	-0.57	-1.32	0.11
Norway	-0.65	-0.78	-0.66	-0.29	0.26	0.86	0.25	1.11	2.05
Portugal	0.56	-0.98	-0.36	-0.82	0.14	0.07	-0.46	-0.49	-0.61
Spain	0.24	-0.47	0.21	-0.55	2.03	-0.71	1.60	-1.20	-0.75
Sweden	-0.84	-0.88	-0.16	-0.29	-0.59	0.31	0.71	1.81	0.18
Switzerland	-0.73	-1.09	1.54	-0.29	0.81	0.99	0.46	-0.68	-0.61
Turkey	3.07	-0.57	-2.73	-2.15	-3.26	-1.70	-1.03	-1.19	-2.40
Bulgaria	0.29	0.67	0.76	-0.82	-0.16	-1.08	-1.18	-0.27	0.11
Czechosolvakia	0.17	1.70	1.21	0.78	0.32	-0.82	-1.11	-0.31	0.33
East Germany	-0.96	1.70	2.03	1.04	-0.34	-0.38	-1.00	0.30	1.33
Hungary	0.16	1.90	0.37	2.64	0.02	-0.78	-1.11	-0.41	1.04
Poland	0.77	1.28	-0.19	-0.02	0.14	-1.19	-1.11	-0.57	0.25
Romania	1.00	0.87	0.44	-0.82	0.32	-1.54	-0.96	-1.22	-1.11
USSR	0.29	0.15	-0.17	-0.82	0.63	-1.50	-1.25	0.52	1.98
Yugoslavia	1.90	0.25	-1.46	0.51	-1.98	-1.43	2.60	-2.16	-1.83

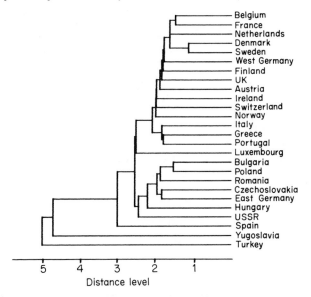

Figure 9.3 Dendogram obtained from a nearest neighbour, hierarchic cluster analysis on employment data from European countries. The scale at the base of the dendogram was added to the output from the program MVSP.

communications. Turkey has extremely high numbers in agriculture and rather low numbers in most other areas.

An alternative analysis was carried out using the cluster analysis option in the program SOLO (BMDP, 1989). This uses the partitioning method described in section 9.2 which starts with arbitrary cluster centres, allocates items to the nearest centre, recalculates the mean values of variables for each group, reallocates individuals to their closest group centres, and so on. The number of clusters is at choice.

For the European employment data solutions with from two to six clusters were obtained. With two clusters, Turkey and Yugoslavia were put into one cluster and the other 24 countries into the second cluster. With six clusters the program chose:

(1)	(2)	(3)	(4)	(5)	(6)
Greece	Ireland	Bulgaria	France	Belgium	Turkey
Portugal	UK	Czechoslovakia	W. Germany	Denmark	Yugoslavia
Spain	Austria	E. Germany	Italy	Netherlands	
Romania	Finland	Hungary	Luxembourg	Norway	
		Poland	Switzerland	Sweden	
		USSR			

This is not the same as the six-cluster solution given by the dendrogram of Fig. 9.3, although there are some similarities. No doubt other algorithms for cluster analysis will give slightly different solutions.

The six clusters produced by the two methods of analysis can be compared with a plot of the countries against the first two principal components of the data (Fig. 6.2). On this basis there is a fair but not perfect agreement: countries in the same cluster tend to have similar values for the first two principal components.

A comparison can also be made between the results of the cluster analyses and a plot of the countries against values for the first two canonical discriminant functions based on a division into EEC, non-EEC western European countries, and eastern European countries (Fig. 8.1). Again there is a fair agreement, with countries in the same cluster tending to have similar values for the canonical discriminant functions.

Example 9.2 *Relationships between canine species*

As a second example, consider the data provided in Table 1.4 for mean mandible measurements of seven canine groups. As has been explained before, these data were originally collected as part of a

Table 9.2 Clusters found at different distance levels for a hierarchic nearest-neighbour cluster analysis (MD = modern dog, GJ = golden jackal, CW = Chinese wolf, IW = Indian wolf, CU = cuon, DI = dingo and PD = prehistoric dog)

Distance		Number of clusters
0	MD, GJ, CW, IW, CU, DI, PD	7
0.72	(MD, PD), GJ, CW, IW, CU, DI	6
1.38	(MD, PD, CU), GJ, CW, IW, DI	5
1.63	(MD, PD, CU), GJ, CW, IW, DI	5
1.68	(MD, PD, CU, DI), GJ, CW, IW	4
1.80	(MD, PD, CU, DI), GJ, CW, IW	4
1.84	(MD, PD, CU, DI), GJ, CW, IW	4
2.07	(MD, PD, CU, DI, GJ), CW, IW	3
2.31	(MD, PD, CU, DI, GJ), (CW, IW)	2
2.37	(MD, PD, CU, DI, GJ, CW, IW)	1

study on the relationship between prehistoric dogs, whose remains have been uncovered in Thailand, and the other six living groups. This question has already been considered in terms of distances between the seven groups in Example 5.1. Table 5.1 shows mandible measurements standardized to have means of zero and standard deviations of one. Table 5.2 shows Euclidean distances between the groups based on these standardized measurements.

With only seven species to cluster it is a simple matter to carry out a nearest-neighbour, hierarchic cluster analysis without using a computer. Thus it can be seen from Table 5.2 that the two most

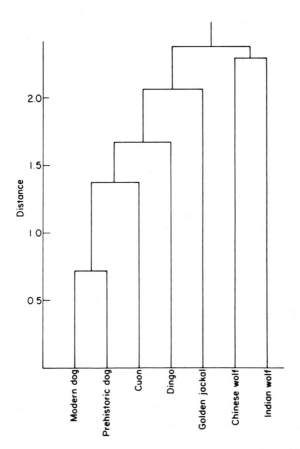

Figure 9.4 Dendogram produced from the clusters shown in Table 9.2.

similar species are the prehistoric dog and the modern dog at a distance of 0.72. These therefore join into a single cluster at that level. The next largest distance is 1.38 between the cuon and the prehistoric dog so that at that level the cuon joins the cluster with the prehistoric and modern dog. The third largest distance is 1.63 between the cuon and modern dog but because these are already in the same cluster this has no effect. Continuing in this way produces the clusters that are shown in Table 9.2. The corresponding dendrogram is shown in Fig. 9.4.

It appears that the prehistoric dog is closely related to the modern Thai dog, with both of these being somewhat related to the cuon and dingo and less closely related to the golden jackal. The Indian and Chinese wolves are closest to each other, but the difference between them is relatively large.

It seems fair to say that in this example the cluster analysis has produced a sensible description of the relationship between the different groups.

9.7 Computational methods and computer programs

Computer programs for cluster analysis are widely available. The larger statistical packages often include a variety of different options for both hierarchic and partitioning methods. However, it is also common to find that only one of these methods is available. For example SOLO (BMDP, 1989) only allows the partitioning approach but MVSP (Kovach, 1993) only allows hierarchic methods.

9.8 Further reading

There are a number of books devoted to cluster analysis that are available. Those by Aldenderfer and Blashfield (1984) and Gordon (1981) are at an introductory level, while more comprehensive accounts are provided by Hartigan (1975) and Romesburg (1984).

An approach to clustering that has not been considered in this chapter involves assuming that the data available come from a mixture of several different populations for which the distributions are of a type that is assumed to be known (e.g. multivariate normal). The clustering problem is then transformed into the problem of estimating, for each of the populations, the parameters of the assumed distribution and the probability that an observation comes from that population. This approach has the merit of moving the clustering problem away from the development of *ad hoc* procedures towards the more usual statistical framework of parameter estimation and model testing. See Everitt and Dunn (1991, p. 113) for an introduction to this method of cluster analysis.

Exercises

1. Table 9.3 shows the abundances of the 25 most abundant plant species on 17 plots from a grazed meadow in Steneryd Nature Reserve in Sweden as measured by Persson (1981) and used for an example by Digby and Kempton (1987). Each value in the table is the sum of cover values in a range from 0 to 5 for nine sample quadrats, so that a value of 45 corresponds to the complete cover by the species being considered. Note that the species are in order from the most abundant (1) to the least abundant (25) and the plots are in the order given by Digby and Kempton (1987, Table 3.2), which corresponds to variation in certain environmental factors such as light and moisture. Carry out cluster analyses to study the relationships between (a) the 17 plots, and (b) the 25 species.

2. Table 9.4 shows a set of data concerning grave goods from a cemetery at Bannadi, northeast Thailand, that were kindly supplied by Professor C.F.W. Higham. These data consist of a record of the presence or absence of 38 different types of article in each of 47 graves, with additional information on whether the body was of an adult male, adult female, or a child. The burials are in the order of richness of the different types of goods (from 0 to 11), and the goods are in the order of the frequency of occurrence (from 1 to 18). Carry out a cluster analysis to study the relationships between the 47 burials. Is there any clustering in terms of the type of body?

Table 9.3 Abundance measures for 25 plant species on 17 plots from a grazed meadow in Steneryd Nature Reserve, Sweden

Species	Plot																
	1	2	3	4	5	6	7	8	9	10	11	12	13	14	15	16	17
1. *Festuca ovina*	38	43	43	30	10	11	20	0	0	5	4	1	1	0	0	0	0
2. *Anemone nemorosa*	0	0	0	4	10	7	21	14	13	19	20	19	6	10	12	14	21
3. *Stallaria holostea*	0	0	0	0	0	6	8	21	39	31	7	12	0	16	11	6	9
4. *Agrostis tenuis*	10	12	19	15	16	9	0	9	28	8	0	4	0	0	0	0	0
5. *Ranunculus ficaria*	0	0	0	0	0	0	0	0	0	0	13	0	0	21	20	21	37
6. *Mercurialis perennis*	0	0	0	0	0	0	0	0	0	0	1	0	0	0	11	45	45
7. *Poa pratenis*	1	0	5	6	2	8	10	15	12	15	4	5	6	7	0	0	0
8. *Rumex acetosa*	0	7	0	10	9	9	3	9	8	9	2	5	5	1	7	0	0
9. *Veronica chamaedrys*	0	0	1	4	6	9	9	9	11	11	6	5	4	1	7	0	0
10. *Dactylis glomerata*	0	0	0	0	0	8	0	14	2	14	3	9	8	7	7	2	1
11. *Fraxinus excelsior* (juv.)	0	0	0	0	0	8	0	0	6	5	4	7	9	8	8	7	6
12. *Saxifraga granulata*	0	5	3	9	12	9	0	1	7	4	5	1	1	1	3	0	0
13. *Deschampsia flexuosa*	0	0	0	0	0	0	30	0	14	3	8	0	3	3	0	0	0
14. *Luzula campestris*	4	10	10	9	7	6	9	0	0	2	1	0	3	0	0	0	0
15. *Plantago lanceolata*	2	9	7	15	13	8	0	0	0	0	0	0	2	0	1	0	0
16. *Festuca rubra*	0	0	0	0	15	6	0	18	1	0	0	0	0	0	0	0	0
17. *Hieracium pilosella*	12	7	16	8	1	6	0	2	1	9	0	7	2	0	0	0	0
18. *Geum urbanum*	0	0	0	0	0	7	0	2	2	1	0	0	0	2	3	8	7
19. *Lathyrus montanus*	0	0	0	0	0	7	9	2	12	6	3	8	9	0	0	0	0
20. *Campanula persicifolia*	0	0	0	0	2	6	3	0	6	5	3	9	3	2	0	0	0
21. *Viola riviniana*	0	0	0	0	0	4	1	4	2	9	6	8	4	1	6	0	0
22. *Hepatica nobilis*	0	0	0	0	0	8	0	0	0	0	2	10	6	0	2	7	0
23. *Achillea millefolium*	1	9	16	9	5	2	0	4	0	0	0	0	0	0	0	0	0
24. *Allium* sp.	0	0	0	0	2	7	0	0	0	3	1	6	8	2	0	0	4
25. *Trifolium repens*	0	0	6	14	19	2	0	0	0	0	0	0	0	0	0	0	0

Table 9.4 Grave goods in burials in the Bannadi cemetery, northeast Thailand. The body types are adult male (1), adult female (2), and child (3)

Burial	Body type	1	2	3	4	5	6	7	8	9	10	11	12	13	14	15	16	17	18	19	20	21	22	23	24	25	26	27	28	29	30	31	32	33	34	35	36	37	38	Sum
B33	3	0	0	0	0	0	0	0	0	0	0	0	0	0	0	0	0	0	0	0	0	0	0	0	0	0	0	0	0	0	0	0	0	0	0	0	0	0	0	0
B9	2	0	0	0	0	0	0	0	0	0	0	0	0	0	0	0	0	0	0	0	0	0	0	0	0	0	0	0	0	0	0	0	0	0	0	0	0	0	0	0
B32	2	0	0	0	0	0	0	0	0	0	0	0	0	0	0	0	0	0	0	0	0	0	0	0	0	0	0	0	0	0	0	0	0	0	0	0	0	0	0	0
B11	1	0	0	0	0	0	0	0	0	0	0	0	0	0	0	0	0	0	0	0	0	0	0	0	0	0	0	0	0	0	0	0	0	0	0	0	0	0	0	0
B28	1	0	0	0	0	0	0	0	0	0	0	0	0	0	0	0	0	0	0	0	0	0	0	0	0	0	0	0	0	0	0	0	0	0	0	0	0	0	0	0
B41	2	0	0	0	0	0	0	0	0	0	0	0	0	0	0	0	0	0	0	0	0	0	0	0	0	0	0	0	0	0	0	0	0	0	0	0	0	0	0	0
B27	2	0	0	0	0	0	0	0	0	0	0	0	0	0	0	0	0	0	0	0	0	0	0	0	0	0	0	0	0	0	0	0	0	0	0	0	0	0	0	0
B24	2	0	0	0	0	0	0	0	0	0	0	0	0	0	0	0	0	0	0	0	0	0	0	0	0	0	0	0	0	0	0	0	0	0	0	0	0	0	0	0
B39	1	0	0	0	0	0	0	0	0	0	0	0	0	0	0	0	0	0	0	0	0	0	0	0	0	0	0	0	0	0	0	0	0	0	0	0	0	0	0	0
B43	2	0	0	0	0	0	0	0	0	0	0	0	0	0	0	0	0	0	0	0	0	0	0	0	0	0	0	0	0	0	0	0	0	0	0	0	0	0	0	0
B20	2	0	0	0	0	0	0	0	0	0	0	0	0	0	0	0	0	0	0	0	0	0	0	0	0	0	0	0	0	0	0	0	0	0	0	0	0	0	0	0
B34	3	0	0	0	0	0	0	0	0	0	0	0	0	0	0	0	0	0	0	0	0	0	0	0	0	0	0	0	0	0	0	0	0	0	0	0	0	0	1	1
B27	1	0	0	0	0	0	0	0	0	0	0	0	0	0	0	0	0	0	0	0	0	0	0	0	0	0	0	0	0	0	0	0	0	0	0	0	0	1	0	1
B37	1	0	0	0	0	0	0	1	0	0	0	0	0	0	0	0	0	0	0	0	0	0	0	0	0	0	0	0	0	0	0	0	0	0	0	0	0	0	0	1
B25	2	0	0	0	0	0	0	0	0	0	0	0	0	0	0	0	0	0	0	0	0	0	0	0	0	0	0	0	0	0	0	1	0	0	0	0	0	0	0	1
B30	2	0	0	0	0	0	0	0	0	0	0	0	0	0	0	0	0	0	0	0	0	0	0	0	0	0	0	0	0	0	1	0	0	0	0	0	0	0	0	1
B21	1	0	0	0	0	0	0	0	0	0	0	0	0	0	0	0	0	0	0	0	0	0	0	0	0	0	0	0	0	0	0	0	0	0	1	0	0	0	0	1
B49	2	0	0	0	0	0	0	0	0	0	0	0	0	0	0	0	0	0	0	0	0	0	0	0	0	0	0	0	0	1	0	0	0	0	0	0	0	0	0	1
B40	2	0	0	0	0	0	0	0	0	0	0	0	0	0	0	0	0	0	0	0	0	0	0	0	0	0	0	0	0	1	0	0	0	0	0	0	0	1	0	2
BT8	2	0	0	0	0	0	0	0	0	0	0	0	0	0	0	0	0	0	0	0	0	0	0	0	0	0	0	0	0	0	0	0	0	0	0	1	1	0	0	2
BT17	2	0	0	0	0	0	0	0	0	0	0	0	0	0	0	0	0	0	0	0	0	0	0	0	0	0	0	0	1	0	0	0	0	0	0	0	1	0	0	2
BT21	2	0	0	0	0	0	0	0	0	0	0	0	0	0	0	0	0	0	0	0	0	0	0	0	0	0	0	0	0	0	0	0	1	0	1	0	0	0	0	2
BT5	1	0	0	0	0	0	0	0	0	0	0	0	0	0	0	0	0	0	0	0	0	0	0	0	0	0	0	0	0	0	0	0	0	0	1	0	1	0	1	3
B14	3	0	0	0	0	0	0	0	0	0	0	0	0	0	0	0	0	0	0	0	0	0	0	0	0	0	0	0	0	0	0	0	0	0	1	0	0	1	1	3

Table 9.4 (*Contd.*)

Burial	Body type	Type of article

Burial	Body type	1	2	3	4	5	6	7	8	9	10	11	12	13	14	15	16	17	18	19	20	21	22	23	24	25	26	27	28	29	30	31	32	33	34	35	36	37	38	Sum
B31	1	0	0	0	0	0	0	0	0	0	0	0	0	0	0	0	0	0	0	0	0	0	1	0	0	0	0	0	0	0	0	0	0	0	1	0	1	0	0	3
B42	1	0	0	0	0	0	0	0	0	0	0	0	0	0	0	0	0	0	0	0	0	0	0	1	0	0	0	0	0	0	0	0	0	0	0	1	0	0	1	3
B44	2	0	0	0	0	0	0	0	0	0	0	0	0	0	0	0	0	0	0	0	0	0	0	0	0	0	0	0	0	0	0	1	0	0	1	0	1	0	0	3
B35	1	0	0	0	0	0	0	0	0	0	0	0	0	0	0	0	0	0	0	0	0	0	0	0	0	0	0	0	0	0	0	0	0	1	0	0	1	0	1	3
BT15	1	0	0	0	0	0	0	0	1	0	0	0	0	0	0	0	0	0	0	0	0	0	0	0	0	0	0	0	0	0	0	0	0	0	1	0	0	1	0	3
B15	3	1	0	0	0	0	0	0	0	0	0	0	0	0	0	0	0	0	0	0	0	0	0	0	0	0	0	0	0	0	0	0	0	0	1	0	1	1	0	4
B45	3	0	0	0	0	0	0	0	0	0	0	0	0	0	0	0	0	0	0	0	0	0	0	1	0	0	0	0	0	0	0	0	0	0	1	0	1	0	1	4
B46	3	0	0	0	0	0	0	0	0	0	0	0	0	0	0	0	0	0	0	0	0	0	0	0	0	0	0	0	0	0	1	0	1	0	0	1	0	1	0	4
B17	1	0	1	0	0	0	0	0	0	0	0	0	0	0	0	0	0	0	0	0	0	0	0	0	0	0	0	0	0	0	0	1	0	0	1	1	0	0	0	4
B10	2	0	0	0	0	0	0	0	0	0	0	0	0	0	0	0	0	0	0	0	0	0	0	0	0	0	0	0	0	0	0	0	0	0	1	1	1	1	0	4
BT16	2	0	0	1	0	0	0	0	0	0	0	0	0	0	0	0	0	0	0	0	0	0	0	0	0	0	0	0	1	0	0	0	0	0	0	1	0	0	1	4
B26	2	0	0	0	0	0	0	0	0	0	0	0	0	0	0	0	0	0	0	0	0	0	0	0	0	0	0	0	1	0	0	0	0	1	0	1	0	1	0	4
B16	1	0	0	0	1	0	0	0	0	0	0	0	0	1	0	0	0	1	0	0	0	0	0	0	0	0	0	0	0	0	0	0	0	0	1	0	0	0	1	5
B29	3	0	0	0	0	0	0	0	0	0	0	0	0	0	0	0	0	0	0	0	0	0	0	0	0	0	1	0	0	0	0	0	1	0	1	0	1	1	0	5
B19	3	0	0	0	0	0	0	0	0	0	0	1	0	0	0	0	0	0	0	0	0	0	0	0	1	0	0	0	0	1	0	0	0	0	1	1	1	0	0	6
B32	2	0	0	0	0	0	0	0	0	0	0	0	0	0	0	0	0	0	0	0	0	0	0	0	1	0	1	0	0	1	0	1	0	0	0	1	0	1	0	6
B38	3	0	0	0	0	0	0	0	0	1	0	0	0	0	0	0	0	0	0	0	0	0	0	0	0	1	0	1	0	1	0	0	0	1	0	0	1	0	1	7
B36	2	0	0	0	0	0	0	0	0	0	0	0	0	0	0	0	0	0	0	0	0	1	0	0	0	0	1	0	0	1	1	0	0	0	1	1	0	0	1	7
B12	2	0	0	0	0	0	0	1	0	0	0	0	0	0	0	0	0	0	0	0	0	0	0	0	0	1	0	1	1	0	0	1	0	0	1	1	0	1	0	8
BT12	1	0	0	0	0	0	0	0	0	0	1	0	0	0	1	0	0	0	0	0	0	0	0	1	0	0	0	0	0	0	1	1	0	1	0	0	1	1	0	8
B47	1	0	0	0	0	1	0	0	0	0	0	0	1	0	0	0	0	0	0	0	0	1	0	0	0	0	0	0	1	0	1	0	1	0	0	1	0	0	1	8
B18	2	0	0	0	0	0	0	0	0	0	0	0	0	0	0	0	0	0	1	1	1	0	0	1	0	1	0	1	0	0	0	0	0	0	1	0	1	1	0	9
B48	2	0	0	0	0	0	0	0	1	1	0	0	0	0	0	1	0	1	0	0	0	0	0	0	0	0	1	0	0	0	1	1	0	0	1	1	1	1	0	11
Sum		1	1	1	1	1	1	1	1	1	1	1	1	1	1	1	1	1	1	1	1	2	2	3	3	3	3	4	4	6	6	7	8	9	12	15	16	18		144

References

Aldenderfer, M.S. and Blashfield, R.K. (1984) *Cluster Analysis*. Sage Publications, Beverly Hills, California.

BMDP (1989) *SOLO User's Guide, Version 3.1*. BMDP Statistical Software, 1440 Sepulveda Boulevard, Los Angeles, California 90025.

Digby, P.G.N. and Kempton, R.A. (1987) *Multivariate Analysis of Ecological Communities*. Chapman and Hall, London.

Everitt, B.S. and Dunn, G. (1991) *Applied Multivariate Data Analysis*. Edward Arnold, London.

Gordon, A.D. (1981) *Classification*. Chapman and Hall, London.

Hartigan, J. (1975) *Clustering Algorithms*. Wiley, New York.

Kovach, W.L. (1993) *MVSP Plus, Version 2.1*. Kovach Computing Services, 85 Nant-y-Felin, Pentraeth, Anglesey LL75 8UY, Wales.

Persson, S. (1981) Ecological indicator values as an aid in the interpretation of ordination diagrams. *Journal of Ecology* **69**, 71–84.

Romesburg, H.C. (1984) *Cluster Analysis for Researchers*. Lifetime Learning Publications, Belmont, California.

CHAPTER TEN

Canonical correlation analysis

10.1 Generalizing a multiple regression analysis

In some sets of multivariate data the variables divide naturally into two groups. A canonical correlation analysis can then be used to investigate the relationships between the two groups. A case in point is the data that are provided in Table 1.3. Here 16 colonies of the butterfly *Euphydryas editha* in California and Oregon are considered. For each colony values are available for four environmental variables and six gene frequencies. An obvious question to be considered is what relationships, if any, exist between the gene frequencies and the environmental variables. One way to investigate this is through a canonical correlation analysis.

Another example was provided by Hotelling (1936) in one of the papers in which he described a canonical correlation analysis for the first time. This example involved the results of tests for reading speed (X_1), reading power (X_2), arithmetic speed (Y_1) and arithmetic power (Y_2) that were given to 140 seventh-grade schoolchildren. The specific question that was addressed was whether or not reading ability (as measured by X_1 and X_2) is related to arithmetic ability (as measured by Y_1 and Y_2). The approach that a canonical correlation analysis takes to answering this question is to search for a linear combination of X_1 and X_2, say

$$U = a_1 X_1 + a_2 X_2,$$

and a linear combination of Y_1 and Y_2, say

$$V = b_1 Y_1 + b_2 Y_2,$$

where these are chosen so that the correlation between U and V is as large as possible. This is somewhat similar to the idea in a principal components analysis except that here a correlation is maximized instead of a variance.

With X_1, X_2, Y_1 and Y_2 standardized to have unit variances, Hotelling found that the best choices for U and V are

$$U = -2.78X_1 + 2.27X_2 \quad \text{and} \quad V = -2.44Y_1 + 1.00Y_2,$$

where these have a correlation of 0.62. It can be seen that U measures the difference between reading power and speed and V measures the difference between arithmetic power and speed. Hence it appears that children with a large difference between X_1 and X_2 tended also to have a large difference between Y_1 and Y_2. It is this aspect of reading and arithmetic that shows most correlation.

In a multiple regression analysis a single variable Y is related to two or more variables $X_1, X_2, \ldots, X_p$ to see how Y is related to the X's (Manly, 1992, Chapter 4). From this point of view, canonical correlation analysis is a generalization of multiple regression in which several Y variables are simultaneously related to several X variables.

In practice more than one pair of canonical variables can be calculated from a set of data. If there are p variables $X_1, X_2, \ldots, X_p$ and q variables $Y_1, Y_2, \ldots, Y_q$ then there can be up to the minimum of p and q pairs of variables. That is to say, linear relationships

$$U_1 = a_{11}X_1 + a_{12}X_2 + \cdots + a_{1p}X_p$$
$$U_2 = a_{21}X_1 + a_{22}X_2 + \cdots + a_{2p}X_p$$
$$\vdots$$
$$U_r = a_{r1}X_1 + a_{r2}X_2 + \cdots + a_{rp}X_p$$

and

$$V_1 = b_{11}Y_1 + b_{12}Y_2 + \cdots + b_{1q}Y_q$$
$$V_2 = b_{21}Y_1 + b_{22}Y_2 + \cdots + b_{2q}Y_q$$
$$\vdots$$
$$V_r = b_{r1}Y_1 + b_{r2}Y_2 + \cdots + b_{rq}Y_q$$

can be established, where r is the smaller of p and q. These relationships are chosen so that the correlation between U_1 and V_1 is a maximum; the correlation between U_2 and V_2 is a maximum, subject to these variables being uncorrelated with U_1 and V_1; the correlation between U_3 and V_3 is a maximum, subject to these variables being uncorrelated with U_1, V_1, U_2 and V_2, and so on. Each of the pairs of canonical variables $(U_1, V_1), (U_2, V_2), \ldots, (U_r, V_r)$ then represents an independent 'dimension' in the relationship between the two sets of variables $(X_1, X_2, \ldots, X_p)$ and $(Y_1, Y_2, \ldots, Y_q)$. The first pair (U_1, V_1) have the highest possible correlation and are therefore the most important; the second pair (U_2, V_2) have the second highest correlation and are therefore the second most important, etc.

10.2 Procedure for a canonical correlation analysis

Assume that the $(p + q) \times (p + q)$ correlation matrix between the variables $X_1, X_2, \ldots, X_p, Y_1, Y_2, \ldots, Y_q$ takes the following form when it is calculated from the sample for which the variables are recorded:

$$
\begin{array}{c}
\begin{array}{cc} X_1 X_2 \ldots X_p & \quad Y_1 Y_2 \ldots Y_q \end{array} \\
\begin{array}{c} X_1 \\ X_2 \\ \vdots \\ X_p \\ Y_1 \\ Y_2 \\ \vdots \\ Y_q \end{array}
\left[
\begin{array}{c|c}
\begin{array}{c} p \times p \text{ matrix} \\ \mathbf{A} \end{array} & \begin{array}{c} p \times q \text{ matrix} \\ \mathbf{C} \end{array} \\
\hline
\begin{array}{c} q \times p \text{ matrix} \\ \mathbf{C}' \end{array} & \begin{array}{c} q \times q \text{ matrix} \\ \mathbf{B} \end{array}
\end{array}
\right].
\end{array}
$$

From this matrix a $q \times q$ matrix $\mathbf{B}^{-1}\mathbf{C}'\mathbf{A}^{-1}\mathbf{C}$ can be calculated, and the eigenvalue problem

$$(\mathbf{B}^{-1}\mathbf{C}'\mathbf{A}^{-1}\mathbf{C} - \lambda\mathbf{I})\mathbf{b} = 0 \qquad (10.1)$$

can be considered. It turns out that the eigenvalues $\lambda_1 > \lambda_2 > \cdots > \lambda_r$ are then the squares of the correlations between the canonical variables and the corresponding eigenvectors, $\mathbf{b}_1, \mathbf{b}_2, \ldots, \mathbf{b}_r$ give the coefficients of the Y variables for the canonical variables. The coefficients of U_i, the ith canonical variable for the X variables, are given by the elements of the vector

$$\mathbf{a}_i = \mathbf{A}^{-1}\mathbf{C}\mathbf{b}_i. \qquad (10.2)$$

In these calculations it is assumed that the original X and Y variables are in a standardized form with means of zero and standard deviations of unity. The coefficients of the canonical variables are for these standardized X and Y variables.

From equations (10.1) and (10.2) the ith pair of canonical variables are calculated as

$$U_i = \mathbf{a}_i'\mathbf{X} = (a_{i1}, a_{i2}, \ldots, a_{ip}) \begin{bmatrix} x_1 \\ x_2 \\ \vdots \\ x_p \end{bmatrix}$$

and

$$V_i = \mathbf{b}_i'\mathbf{Y} = (b_{i1}, b_{i2}, \ldots, b_{iq}) \begin{bmatrix} y_1 \\ y_2 \\ \vdots \\ y_q \end{bmatrix}$$

where $\mathbf{X}$ and $\mathbf{Y}$ are vectors of standardized data values. As they stand, U_i and V_i will have variances that depend upon the scaling adopted for the eigenvector $\mathbf{b}_i$. However, it is a simple matter to calculate the standard deviation of U_i for the data and divide the a_{ij} values by this standard deviation. This produces a scaled canonical variable U_i with unit variance. Similarly, if the b_{ij} values are divided by the standard deviation of V_i then this produces a scaled V_i with unit variance.

This form of standardization of the canonical variables is not essential because the correlation between U_i and V_i is not affected by scaling. However, it may be useful when it comes to examining the numerical values of canonical variables for the individuals for which data are available.

10.3 Tests of significance

An approximate test for a relationship between the X variables as a whole and the Y variables as a whole was proposed by Bartlett (1947) for the situation where the data are a random sample from a multivariate normal distribution. This involves calculating the statistic

$$X^2 = -\{n - \tfrac{1}{2}(p + q + 3)\} \sum_{i=1}^{r} \log_e(1 - \lambda_i), \qquad (10.3)$$

where n is the number of cases for which data are available. The statistic can be compared with the percentage points of the chi-squared distribution with pq degrees of freedom, and a significantly large value provides evidence that at least one of the r canonical correlations is significant. A non-significant result indicates that even the largest canonical correlation can be accounted for by sampling variation only.

It is sometimes suggested that this test can be extended to allow the importance of each of the canonical correlations to be tested. Common suggestions are to:

1. compare the ith contributions, $-\{n - \tfrac{1}{2}(p + q + 3)\} \log_e(1 - \lambda_i)$, to the right-hand side of equation (10.3) to the percentage points of the chi-squared distribution with $p + q - 2i + 1$ degrees of freedom; or

2. compare the sum of the $(i + 1)$th to the rth contributions to the sum on the right-hand side of equation (10.3) to the percentage points of the chi-squared distribution with $(p - i)(q - i)$ degrees of freedom.

Here (1) is assumed to be testing the ith canonical correlation directly, whereas (2) is assumed to be testing for the significance of the $(i + 1)$th to rth canonical correlations as a whole.

The reason why these tests are *not* reliable is essentially the same as has already been discussed in section 8.4 for a related test used with discriminant function analysis: the ith largest sample canonical correlation may in fact have arisen from a population canonical correlation that is not the ith largest. Hence the association between the r contributions to the right-hand side of equation (10.3) and the r population canonical correlations is blurred. See Harris (1985, p. 211) for a further discussion about this matter.

There are some modifications of the test statistic X^2 that are sometimes proposed to improve the chi-squared approximation for the distribution of this statistic when the null hypothesis holds and the sample size is small. These will not be considered here.

10.4 Interpreting canonical variates

If

$$U_i = a_{i1}X_1 + a_{i2}X_2 + \cdots + a_{ip}X_p$$

and

$$V_i = b_{i1}Y_1 + b_{i2}Y_2 + \cdots + b_{iq}Y_q$$

then it seems that U_i can be described in terms of the X variables with large coefficients a_{ij} and V_i can be described in terms of the Y variables with large coefficients b_{ij}. 'Large' here means, of course, large positive or large negative.

Unfortunately, correlations between the X and Y variables can upset this interpretation process. For example, it can happen that a_{i1} is positive and yet the simple correlation between U_i and X_1 is actually negative. This apparent contradiction can come about because X_1 is highly correlated with one or more of the other X variables and part of the effect of X_1 is being accounted for by the coefficients of these other X variables. In fact, if one of the X variables is almost a linear combination of the other X variables then there will be an infinite variety of linear combinations of the X variables, some of them with very different a_{ij} values, that give virtually the same U_i values. The same can be said about linear combinations of Y variables.

The interpretation problems that arise with highly correlated X variables or Y variables should be familiar to users of multiple regression analysis. Exactly the same problems arise with the estimation of regression coefficients.

Actually, a fair comment seems to be that if the X or Y variables are highly correlated then there can be no way of disentangling their contributions to canonical variables. However, no doubt people will continue to try to make interpretations under these circumstances.

Some authors have suggested that it is better to describe canonical variables by looking at their correlations with the X and Y variables rather than the coefficients a_{ij} and b_{ij}. For example, if U_i is highly positively correlated with X_1 then U_i can be considered to reflect X_1 to a large extent. Similarly, if V_i is highly negatively correlated with Y_1 then V_i can be considered to reflect the opposite of Y_1 to a large extent. This approach does at least have the merit of bringing out all of the variables to which the canonical variables seem to be related.

Example 10.1 Environmental and genetic correlations
for colonies of Euphydryas editha

The data in Table 1.3 can be used to illustrate the procedure for a canonical correlation analysis. Here there are 16 colonies of the butterfly *Euphydryas editha* in California and Oregon. These vary with respect to four environmental variables (altitude, annual precipitation, annual maximum temperature, and annual minimum temperature) and six genetic variables (percentages of six phosphoglucose-isomerase genes as determined by electrophoresis). Any significant relationships between the environmental and genetic variables are interesting because they may indicate the adaption of *E. editha* to the environment.

For a canonical correlation analysis the environmental variables have been treated as the X variables and the gene frequencies as the Y variables. However, the six gene frequencies have not been used as shown in Table 1.3 because they add up to 100%, which allows different linear combinations of these variables to have the same correlation with a combination of the Y variables. To see this, suppose that the first pair of canonical variables are U_1 and V_1, where $U_1 = a_{11}X_1 + a_{12}X_2 + \cdots + a_{1p}X_p$. Then U_1 can be rewritten by replacing X_1 by 100 minus the sum of the other variables to give

$$U_1 = 100a_{11} + (a_{12} - a_{11})X_2 + \cdots + (a_{1p} - a_{11})X_p.$$

This means that the correlation between U_1 and V_1 is the same as that between $(a_{12} - a_{11})X_2 + \cdots + (a_{1p} - a_{11})X_p$ and V_1, because the constant $100a_{11}$ in the second linear combination has no effect on the correlation. Thus two linear combinations of the X variables with very different coefficients can serve just as well for the canonical variable. In fact, it can be shown that an infinite number of different linear combinations of the X variables will serve just as well, and the same holds true for linear combinations of standardized X variables.

This problem is overcome by removing one of the gene frequencies from the analysis. Here the 1.30 gene frequency was omitted. The data were also further modified by combining the frequencies for the 0.40 and 0.60 mobility genes. Thus the X variables being considered are X_1 = altitude, X_2 = annual precipitation, X_3 = annual maximum temperature and X_4 = annual minimum temperature, while the Y variables are Y_1 = frequency of 0.40 and 0.60 mobility genes, Y_2 = frequency of 0.80 mobility genes, Y_3 = frequency of 1.00 mobility genes and Y_4 = frequency of 1.16 mobility genes. Following the development in section 10.2 it is the standardized values of the variables that have been analysed so that for the remainder of this example X_i and Y_i refer to the standardized X and Y variables.

The correlation matrix for the eight variables is shown in Table 10.1, partitioned into the submatrices **A**, **B**, **C** and **C'**, as described in section 10.2. The eigenvalues obtained from equation (10.1) are 0.7425, 0.2049, 0.1425 and 0.0069. Taking square roots gives the corresponding canonical correlations of 0.8617, 0.4527, 0.3775 and 0.0833, respectively, and the canonical variables are found to be as follows:

$$U_1 = -0.09X_1 - 0.29X_2 + 0.48X_3 + 0.29X_4,$$
$$V_1 = +0.54Y_1 + 0.42Y_2 - 0.10Y_3 + 0.82Y_4,$$

$$U_2 = +2.31X_1 - 0.73X_2 + 0.45X_3 + 1.27X_4,$$
$$V_2 = -1.66Y_1 - 2.20Y_2 - 3.71Y_3 + 2.77Y_4,$$

$$U_3 = +3.02X_1 + 1.33X_2 + 0.57X_3 + 3.58X_4,$$
$$V_3 = -3.56Y_1 - 1.35Y_2 - 3.86Y_3 - 2.86Y_4,$$

$$U_4 = +1.43X_1 + 0.26X_2 + 1.72X_3 - 0.03X_4,$$
$$V_4 = +0.60Y_1 - 1.44Y_2 - 0.58Y_3 + 0.58Y_4.$$

There are four canonical correlations because this is the minimum of the number of X variables and the number of Y variables (which both happen to be equal to four).

Although the canonical correlations are quite large, they are not significantly so according to Bartlett's test, because of the small sample size. It is found that $\lambda = 18.34$ with 16 degrees of freedom, where the probability of a value this large from the chi-squared distribution is about 0.30.

Laying aside the lack of significance, it is interesting to see what interpretation can be given to the first pair of canonical variables. From the equation for U_1 it can be seen that this is mainly a contrast between X_3 (maximum temperature) and X_4 (minimum temperature)

Table 10.1 Correlation matrix for variables measured on colonies of *Euphydryas editha*, partitioned into **A**, **B**, **C** and **C'** submatrices

	X_1	X_2	X_3	X_4	Y_1	Y_2	Y_3	Y_4
X_1	1.0000	0.5675	-0.8277	-0.9360	-0.2012	-0.5729	0.7270	-0.4577
X_2	0.5675	1.0000	-0.4787	-0.7046	-0.4684	-0.5498	0.6990	-0.1380
X_3	-0.8277	-0.4787	1.0000	0.7191	0.2242	0.5358	-0.7173	0.4383
X_4	-0.9360	-0.7046	0.7191	1.0000	0.2456	0.5933	-0.7590	0.4122
Y_1	-0.2012	-0.4684	0.2242	0.2456	1.0000	0.6377	-0.5610	-0.5843
Y_2	-0.5729	-0.5498	0.5358	0.5933	0.6377	1.0000	-0.8235	-0.1267
Y_3	0.7270	0.6990	-0.7173	-0.7590	-0.5610	-0.8235	1.0000	-0.2638
Y_4	-0.4577	-0.1380	0.4383	0.4122	-0.5843	-0.1267	-0.2638	1.0000

Partition labels: **A** (upper-left), **C'** (upper-right), **C** (lower-left), **B** (lower-right).

on the one hand, and X_2 (precipitation) on the other. For V_1 there are moderate to large positive coefficients for Y_1 (0.40 and 0.60 mobility), Y_2 (0.80 mobility) and Y_4 (1.16 mobility), and a small negative coefficient for Y_3 (1.00 mobility). It appears that the 0.40, 0.60, 0.80 and 1.16 mobility genes tend to be frequent in colonies with high temperatures and low precipitation.

The correlations between the environmental variables and U_1 are shown below:

	Altitude	Precipitation	Maximum temperature	Minimum temperature
U_1	-0.92	-0.77	0.90	0.92

This suggests that U_1 is best interpreted as a measure of high temperatures and low altitude and precipitation. The correlations between V_1 and the gene frequencies are:

	Mobility 0.40/0.60	Mobility 0.80	Mobility 1.00	Mobility 1.16
V_1	0.38	0.74	-0.96	0.49

In this case V_1 comes out clearly as indicating a lack of mobility 1.00 genes.

The interpretations of U_1 and V_1 are not the same when made on the basis of the coefficients of the canonical functions as they are on the basis of correlations. For U_1 the difference is not great and only concerns the status of altitude, but for V_1 the importance of the mobility 1.00 genes is very different. On the whole, the interpretations based on correlations seem best and correspond with what is seen in the data. For example, colony GL has the highest altitude, high precipitation, the lowest temperatures and the highest frequency of 1.00 mobility genes. This compares with colony UO with a low altitude, low precipitation, high temperature and the lowest frequency of mobility 1.00 genes. However, as mentioned in the previous section, there are real problems with interpreting canonical variables when the variables that they are constructed from have high correlations. Table 10.1 shows that this is indeed the case with this example.

Figure 10.1 shows a plot of the values of V_1 against the values of U_1. It is immediately clear that the colony labelled DP is somewhat unusual compared to the other colonies because the value of V_1 is not similar to that for other colonies with about the same values for U_1. From the interpretations given for U_1 and V_1 it would seem that the frequency of mobility 1.00 genes is unusually high for a

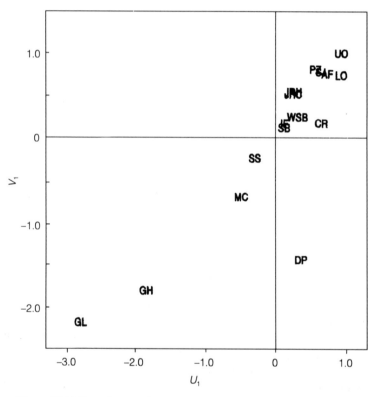

Figure 10.1 Plot of V_1 against U_1 for 16 colonies of *Euphydryas editha*.

colony with this environment. Inspection of the data in Table 1.3 shows immediately that this is the case.

Example 10.2 Soil and vegetation variables in Belize

For an example with·a larger data set, consider part of the data collected by Green (1973) for a study on the factors influencing the location of prehistoric Maya habitation sites in the Corozal district of Belize in Central America. Table 10.2 shows four soil variables and four vegetation variables recorded for 2.5 km by 2.5 km squares. Here canonical correlation analysis is used to study the relationship between these two groups of variables.

The soil variables are $X_1 = \%$ soil with constant lime enrichment, $X_2 = \%$ meadow soil with calcium groundwater, $X_3 = \%$ soil with coral bedrock under conditions of constant lime enrichment and $X_4 = \%$ alluvial and organic soils adjacent to rivers and saline organic soil at the coast. The vegetation variables are $Y_1 = \%$ deciduous seasonal broadleaf forest, $Y_2 = \%$ high and low marsh forest, herbaceous marsh and swamp, $Y_3 = \%$ cohune palm forest and $Y_4 = \%$ mixed forest. The percentages do not add to 100 for all of the squares so there is no need to remove any variables before starting the analysis.

There are four canonical correlations (the minimum of the number of X variables and the number of Y variables) and they are found to be 0.762, 0.566, 0.243 and 0.122. The statistic X^2 is found to equal 193.63 with 16 degrees of freedom from equation (10.3), which is very highly significantly large when compared with the percentage points of the chi-squared distribution. Therefore there is apparently very strong evidence that the soil and vegetation variables are related. However, the original data are clearly not normally distributed so this result must be treated with some reservations.

The canonical variables are found to be as follows:

$$U_1 = + 1.34X_1 + 0.34X_2 + 1.13X_3 + 0.59X_4,$$
$$V_1 = + 1.71Y_1 + 1.07Y_2 + 0.22Y_3 + 0.52Y_4,$$
$$U_2 = + 0.41X_1 + 0.90X_2 + 0.23X_3 + 0.89X_4,$$
$$V_2 = + 0.64Y_1 + 1.47Y_2 + 0.27Y_3 + 0.28Y_4,$$
$$U_3 = + 0.44X_1 - 0.51X_2 + 0.18X_3 + 0.93X_4,$$

Table 10.2 Soil and vegetation variables for 151 2.5 km by 2.5 km squares in the Corozal region of Belize in Central America (X_1 = % soil with constant lime enrichment, X_2 = % meadow soil with calcium groundwater, X_3 = % soil with coral bedrock under conditions of constant lime enrichment, X_4 = % alluvial and organic soils adjacent to rivers and saline organic soil at the coast, Y_1 = % deciduous seasonal broadleaf forest, Y_2 = % high and low marsh forest, herbaceous marsh and swamp, Y_3 = % cohune palm forest and Y_4 = % mixed forest)

Case	X_1	X_2	X_3	X_4	Y_1	Y_2	Y_3	Y_4	Case	X_1	X_2	X_3	X_4	Y_1	Y_2	Y_3	Y_4
1	40	30	0	30	0	25	0	0	51	50	0	0	0	40	0	0	0
2	20	0	0	10	10	90	0	0	52	30	30	0	20	30	60	0	0
3	5	0	0	50	20	50	0	0	53	20	20	0	40	0	100	0	0
4	30	0	0	30	0	60	0	0	54	20	80	0	0	0	100	0	0
5	40	20	0	20	0	95	0	0	55	0	10	0	60	0	75	0	0
6	60	20	0	5	0	100	0	0	56	0	50	0	30	0	75	0	0
7	90	0	0	10	0	100	0	0	57	50	50	0	0	30	70	0	0
8	100	0	0	0	20	80	0	0	58	0	0	0	60	0	60	0	0
9	0	0	0	10	40	60	0	0	59	20	20	0	60	0	100	0	0
10	15	0	0	20	25	10	0	0	60	90	10	0	0	70	30	0	0
11	20	0	0	10	5	50	0	0	61	100	0	0	0	100	0	0	0
12	0	0	0	50	5	60	0	0	62	15	15	0	30	0	40	0	0
13	10	0	0	30	30	60	0	0	63	100	0	0	0	25	75	0	0
14	40	0	0	20	50	10	0	0	64	95	0	0	5	90	10	0	0
15	10	0	0	40	80	20	0	0	65	95	0	0	5	90	10	0	0
16	60	0	0	0	100	0	0	0	66	60	40	0	0	50	50	0	0
17	45	0	0	0	5	60	0	0	67	30	60	10	10	50	40	0	0
18	100	0	0	0	100	0	0	0	68	50	0	50	50	100	0	0	0
19	20	0	0	0	20	0	0	0	69	60	30	0	10	60	40	0	0
20	0	0	0	60	0	50	0	0	70	90	8	0	2	80	20	0	0
21	0	0	0	80	0	75	0	0	71	30	30	30	40	60	40	0	0

72	0	0	25	75	33	33	33	33
73	0	0	100	0	40		10	20
74	0	0	60	40	50		0	50
75	0	0	50	50	12		12	75
76	0	0	60	40	25		0	75
77	0	0	100	0	50		10	30
78	0	0	95	5	30		0	50
79	0	0	40	60	0		0	100
80	0	0	80	20	50		0	50
81	0	0	100	0	90		30	10
82	0	0	85	0	20		20	30
83	0	0	75	50	20		0	20
84	0	0	25	30	0		0	90
85	0	0	5	20	50		30	30
86	0	0	80	50	10		30	20
87	0	0	50	70	0		0	50
88	0	0	10	50	25		10	80
89	0	0	0	80	0		0	80
90	0	0	15	75	0		0	60
91	0	0	0	75	0		30	50
92	0	0	0	85	0		20	70
93	0	0	15	40	0		0	100
94	0	0	60	50	60		0	60
95	0	0	50	100	30		0	80
96	0	0	0	95	35		20	100
97	0	0	5	0			0	100
98	0	0	50	0				0
99	40	0	60	20	60			30
100	0	0	30	20	30			15

22	0	0	50	0	50	0	0	0
23	0	0	100	0	60	0	10	30
24	0	0	50	0	50	0	0	0
25	0	0	100	0	30	0	20	50
26	0	0	100	10	80	0	15	5
27	0	0	90	50	0	0	40	60
28	0	0	50	90	0	0	40	60
29	0	0	10	0	20	0	5	94
30	0	0	100	25	0	0	0	80
31	0	0	75	75	0	50	50	50
32	0	0	25	10	0	75	40	10
33	0	0	90	15	0	0	12	12
34	0	0	85	80	0	10	50	50
35	0	0	20	100	0	100	40	50
36	0	0	0	100	0	100	0	0
37	0	0	0	0	0	0	0	0
38	0	0	50	50	0	20	30	70
39	0	0	50	50	0	100	40	40
40	0	0	0	100	0	50	0	0
41	0	0	20	100	0	0	25	25
42	0	0	0	80	20	0	40	40
43	0	0	0	100	10	0	0	90
44	0	0	0	100	0	0	0	100
45	0	0	10	90	0	0	0	100
46	0	0	0	100	90	0	0	10
47	0	0	0	80	20	0	0	80
48	0	0	0	0	30	0	0	60
49	0	0	30	0	0	0	0	40
50	0	0	0	100	0	0	0	50

Table 10.2 (*Contd.*)

Case	X_1	X_2	X_3	X_4	Y_1	Y_2	Y_3	Y_4
101	40	0	0	45	70	20	0	0
102	30	0	0	45	20	40	0	20
103	60	10	0	30	10	65	5	20
104	40	20	0	40	0	25	0	75
105	100	0	0	0	70	0	0	30
106	100	0	0	0	40	60	0	0
107	80	10	0	10	40	60	0	0
108	90	0	0	10	10	0	0	90
109	100	0	0	0	20	10	0	70
110	30	50	0	20	10	90	0	0
111	60	40	0	0	50	50	0	0
112	100	0	0	0	80	10	0	10
113	60	0	0	40	60	10	30	0
114	50	50	0	0	0	100	0	0
115	60	30	0	10	25	75	0	0
116	40	0	0	60	30	20	50	0
117	30	0	0	70	0	50	50	0
118	50	20	0	30	0	100	0	0
119	50	50	0	0	25	75	0	0
120	90	10	0	0	50	50	0	0
121	100	0	0	0	60	40	0	0
122	50	0	0	50	70	30	0	0
123	10	10	0	80	0	100	0	0
124	50	50	0	0	30	70	0	0
125	75	0	0	25	80	20	0	0
126	40	0	0	60	0	100	0	0

Case	X_1	X_2	X_3	X_4	Y_1	Y_2	Y_3	Y_4
127	90	10	0	10	75	25	0	0
128	45	45	0	55	30	70	0	0
129	20	35	0	80	10	90	0	0
130	80	0	0	20	70	30	0	0
131	100	0	0	0	90	0	0	0
132	75	0	0	25	50	50	0	0
133	60	5	0	40	50	50	0	0
134	40	0	0	60	60	40	0	0
135	60	0	0	40	70	15	0	0
136	90	10	0	10	75	25	0	0
137	50	0	5	0	30	20	0	0
138	70	0	30	0	70	30	0	0
139	60	0	40	0	100	0	0	0
140	50	0	0	0	50	0	0	0
141	30	0	50	0	60	40	0	0
142	5	0	95	0	80	20	0	0
143	10	0	90	0	70	30	0	0
144	50	0	0	0	15	30	0	0
145	20	0	80	0	50	50	0	0
146	0	0	100	0	90	10	0	0
147	0	0	100	0	75	25	0	0
148	90	0	10	0	60	30	10	0
149	0	0	100	0	80	10	10	0
150	0	0	100	0	60	40	0	0
151	0	40	60	40	50	50	0	0

$$V_3 = -0.18Y_1 - 0.24Y_2 + 0.93Y_3 + 0.22Y_4,$$
$$U_4 = -0.44X_1 - 0.02X_2 + 0.72X_3 + 0.15X_4,$$
$$V_4 = +0.12Y_1 + 0.01Y_2 + 0.26Y_3 - 0.93Y_4.$$

In fact, the linear combinations given here for U_1, V_1, U_2 and V_2 are not the ones that were output by the program used to do the calculations because the output linear combinations all had negative coefficients for X and Y variables. A switch in sign is justified because the correlation between $-U_i$ and $-V_i$ is the same as that between U_i and V_i. Hence $-U_i$ and $-V_i$ will serve as well as U_i and V_i as the ith canonical variables. Note, however, that switching signs for U_1, V_1, U_2 and V_2 has changed the signs of the correlations between these canonical variables and the X and Y variables, as shown in Table 10.3.

By considering the correlations shown in Table 10.3 (particularly those outside the range -0.5 to $+0.5$) it appears that the canonical variables can be described as mainly measuring the following dimensions:

U_1: the presence of soil types 1 (soil with constant lime enrichment) and 3 (soil with coral bedrock under conditions of constant lime enrichment) and the absence of soil type 4 (alluvial and organic soils adjacent to rivers and saline organic soil at the coast);

V_1: the presence of vegetation type 1 (deciduous seasonal broadleaf forest);

U_2: the presence of soil types 2 (meadow soil with calcium groundwater) and 4;

V_2: the presence of vegetation type 2 (high and low marsh forest, herbaceous marsh and swamp) and the absence of vegetation type 1;

U_3: the presence of soil type 4 and the absence of soil type 2;

V_3: the pesence of vegetation type 3 (cohune palm forest);

U_4: the presence of soil type 3 and the absence of soil type 1;

V_4: the absence of vegetation type 4 (mixed forest).

It appears, therefore, that the most important relationships between the soil and vegetation variables, that are described by the first two pairs of canonical variables, are (a) the presence of soil types 1 and

Table 10.3 Correlations between canonical variables and the X and Y variables

	U_1	U_2	U_3	U_4		V_1	V_2	V_3	V_4
X_1	0.55	−0.23	−0.00	−0.80	Y_1	0.77	−0.58	−0.08	0.24
X_2	−0.02	0.73	−0.68	−0.04	Y_2	−0.02	0.73	−0.68	−0.04
X_3	0.41	−0.24	−0.18	0.86	Y_3	0.41	−0.24	−0.18	0.86
X_4	−0.35	0.55	0.74	0.19	Y_4	−0.35	0.55	0.74	0•19

3 and the absence of soil type 4 is associated with the presence of vegetation type 1, and (b) the presence of soil types 2 and 4 is associated with the presence of vegetation type 2 and the absence of vegetation type 1.

It is instructive to examine a draftsman's plot of the canonical variables and the case numbers, as shown in Fig. 10.2. The strong correlations between U_1 and V_1 and between U_2 and V_2 are apparent, as might be expected. Perhaps the most striking thing shown by the plots is the unusual distributions of V_3 and V_4. Most values are very similar, at about -0.2 for V_3 and about $+0.2$ for V_4. However, there are extreme values for some cases (observations) between 100 and 120. Inspection of the data in Table 10.2 shows that these extreme cases are for squares where vegetation types 3 and 4 were present, which makes perfect sense from the definition of V_3 and V_4.

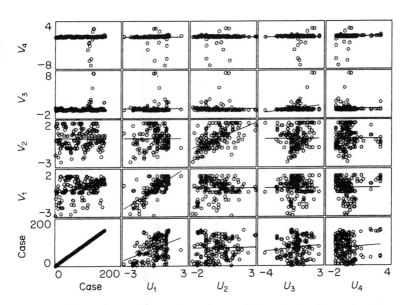

Figure 10.2 Draftsman's plot of canonical variables obtained from the data on soil and vegetation variables for 2.5 km squares in Belize. The line shown in each plot is for the regression of the vertical variable against the horizontal variable.

Table 10.4 Sources of protein (RM, red meat; WM, white meat; EGG, eggs; MLK, milk; FSH, fish; CRL, cereals; SCH, starchy foods; PNO, pulses, nuts and oilseed; F&V, fruit and vegetables) and percentages employed in different industry groups (AGR, agriculture; MIN, mining; MAN, manufacturing; PS, power supplies; CON, construction; SER, service industries; FIN, finance; SPS, social and personal services; TC, transport and communications) for countries in Europe

Country	Protein consumption (g/(person day))									Percentages employed in								
	RM	WM	EGG	MLK	FSH	CRL	SCH	PNO	F&V	AGR	MIN	MAN	PS	CON	SER	FIN	SPS	TC
Austria	9	14	4	20	2	28	4	1	4	12.7	1.1	30.2	1.4	9.0	16.8	4.9	16.8	7.0
Belgium	14	9	4	18	5	27	6	2	4	3.3	0.9	27.6	0.9	8.2	19.1	6.2	26.6	7.2
Bulgaria	8	6	2	8	1	57	1	4	4	23.6	1.9	32.3	0.6	7.9	8.0	0.7	18.2	6.7
Czechoslovakia	10	11	3	13	2	34	5	1	4	16.5	2.9	35.5	1.2	8.7	9.2	0.9	17.9	7.0
Denmark	11	11	4	25	10	22	5	1	2	9.2	0.1	21.8	0.6	8.3	14.6	6.5	32.2	7.1
E. Germany	8	12	4	11	5	25	7	1	4	4.2	2.9	41.2	1.3	7.6	11.2	1.2	22.1	8.4
Finland	10	5	3	34	6	26	5	1	1	13.0	0.4	25.9	1.3	7.4	14.7	5.5	24.3	7.6
France	18	10	3	20	6	28	5	2	7	10.8	0.8	27.5	0.9	8.9	16.8	6.0	22.6	5.7

Greece	10	3	3	18	6	42	2	8	7	41.4	0.6	17.6	0.6	8.1	11.5	2.4	11.0	6.7
Hungary	5	12	3	10	0	40	4	5	4	21.7	3.1	29.6	1.9	8.2	9.4	0.9	17.2	8.0
Ireland	14	10	5	26	2	24	6	2	3	23.2	1.0	20.7	1.3	7.5	16.8	2.8	20.8	6.1
Italy	9	5	3	14	3	37	2	4	7	15.9	0.6	27.6	0.5	10.0	18.1	1.6	20.1	5.7
Netherlands	10	14	4	23	3	22	4	2	4	6.3	0.1	22.5	1.0	9.9	18.0	6.8	28.5	6.8
Norway	9	5	3	23	10	23	5	2	3	9.0	0.5	22.4	0.8	8.6	16.9	4.7	27.6	9.4
Poland	7	10	3	19	3	36	6	2	7	31.1	2.5	25.7	0.9	8.4	7.5	0.9	16.1	6.9
Portugal	6	4	1	5	14	27	6	5	8	27.8	0.3	24.5	0.6	8.4	13.3	2.7	16.7	5.7
Romania	6	6	2	11	1	50	3	5	3	34.7	2.1	30.1	0.6	8.7	5.9	1.3	11.7	5.0
Spain	7	3	3	9	7	29	6	6	7	22.9	0.8	28.5	0.7	11.5	9.7	8.5	11.8	5.5
Sweden	10	8	4	25	8	20	4	1	2	6.1	0.4	25.9	0.8	7.2	14.4	6.0	32.4	6.8
Switzerland	13	10	3	24	2	26	3	2	5	7.7	0.2	37.8	0.8	9.5	17.5	5.3	15.4	5.7
UK	17	6	5	21	4	24	5	3	3	2.7	1.4	30.2	1.4	6.9	16.9	5.7	28.3	6.4
USSR	9	5	2	17	3	44	6	3	3	23.7	1.4	25.8	0.6	9.2	6.1	0.5	23.6	9.3
W. Germany	11	13	4	19	3	19	5	2	4	6.7	1.3	35.8	0.9	7.3	14.4	5.0	22.3	6.1
Yugoslavia	4	5	1	10	1	56	3	6	3	48.7	1.5	16.8	1.1	4.9	6.4	11.3	5.3	4.0

Before leaving this example it is appropriate to make mention of a potential problem that has not been addressed. This concerns the potential spatial correlation in the data for squares that are close in space, and particularly those that are adjacent. If such correlation exists so that, for example, neighbouring squares tend to have the same soil and vegetation characteristics, then the data do not provide 151 independent observations. In effect, the data set will be equivalent to independent data from some smaller number of squares. The effect of this will appear mainly in the test for the significance of the canonical correlations as a whole, with a tendency for these correlations to appear to be more significant than they really are.

The same problem also potentially exists with the previous data on colonies of the butterfly *Euphydryas editha* because some of the colonies were quite close in space. Indeed, it is a potential problem whenever observations are taken in different locations in space. The way to avoid the problem is to ensure that observations are taken sufficiently far apart that they are independent or close to independent, although this is often easier said than done. There are methods available for allowing for spatial correlation in data, but these are beyond the scope of this book.

10.5 Computational methods and computer programs

Computer programs for canonical correlation analysis are not as widely available as programs for the multivariate analyses considered in earlier chapters, although it is an option in the larger packages such as BMDP (Dixon, 1990), SAS (1985) and SPSS (1990). The calculations for the examples in this chapter were done using a general package for statistical analysis called SURVO (Mustonen, 1992) that operates similarly to a spreadsheet. However, this is not widely available at the present time.

10.6 Further reading

See Harris (1985) for a fuller discussion about the theory of canonical correlation analysis. Another useful reference is a book by Giffins (1985) on canonical correlation analysis for applications in ecology. About half of this text is devoted to theory and the remainder to specific examples on plants.

Exercise

Table 10.4 shows the result of combining the data in Tables 1.5 and 6.7 on sources of protein and employment patterns for European countries, for the 24 countries where this is possible. Use canonical correlation analysis to investigate the relationship, if any, between the nature of the employment in a country and the type of food that is used for protein.

References

Bartlett, M.S. (1947) The general canonical correlation distribution. *Annals of Mathematical Statistics* **18**, 1–17.

Dixon, W.J. (ed.) (1990) *BMDP Statistical Software Manual.* University of California Press, Berkeley.

Giffins, R. (1985) *Canonical Analysis: A Review with Applications in Ecology.* Springer-Verlag, Berlin.

Green, E.L. (1973) Location analysis of prehistoric Maya sites in British Honduras. *American Antiquity* **38**, 279–93.

Harris, R.J. (1985) *A Primer of Multivariate Statistics.* Academic Press, Orlando.

Hotelling, H. (1936) Relations between two sets of variables. *Biometrika* **28**, 321–77.

Manly, B.F.J. (1992) *The Design and Analysis of Research Studies.* Cambridge University Press, Cambridge.

Mustonen, S. (1992) *SURVO, An Integrated Environment for Statistical Computing and Related Areas.* Survo Systems Ltd, Helsinki, Finland.

SAS (1985) *SAS User's Guide: Statistics.* SAS Institute, Cary, North Carolina 27511.

SPSS (1990) *SPSS Reference Guide.* SPSS Inc., 444 N. Michigan Avenue, Chicago Illinois 60611.

CHAPTER ELEVEN

Multidimensional scaling

11.1 Constructing a 'map' from a distance matrix

Multidimensional scaling is a technique that is designed to construct a 'map' showing the relationships between a number of objects, given only a table of distances between them. The 'map' can be in one dimension (if the objects fall on a line), in two dimensions (if the objects lie on a plane), in three dimensions (if the objects can be represented by points in space), or in a higher number of dimensions (in which case in immediate geometrical representation is not possible).

The fact that it may be possible to construct a map from a table of distances can be seen by considering the example of four objects A, B, C and D shown in Fig. 11.1. Here the distances apart are given by the array:

	A	B	C	D
A	0	6.0	6.0	2.5
B	6.0	0	9.5	7.8
C	6.0	9.5	0	3.5
D	2.5	7.8	3.5	0

For example, the distance from A to B, which is the same as the distance from B to A, is 6.0. The distance of objects to themselves is, of course, 0. It seems plausible that the map can be reconstructed from the array of distances. A moment's reflection may indicate, however, that a mirror image of the map as shown in Fig. 11.2 will have the same array of distances between objects. Consequently, it seems clear that a recovery of the original map will be subject to a possible reversal of this type.

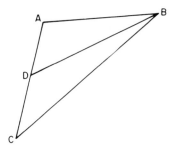

Figure 11.1 A map of the relationship between four objects.

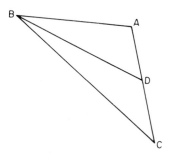

Figure 11.2 A mirror image of the map shown in Fig. 11.1 for which the distances between objects are the same.

It is also apparent that if more than three objects are involved then they may not lie on a plane. In that case their distance matrix will implicitly contain this information. For example, the distance array

	A	B	C	D
A	0	1	$\sqrt{2}$	$\sqrt{2}$
B	1	0	1	1
C	$\sqrt{2}$	1	0	$\sqrt{2}$
D	$\sqrt{2}$	1	$\sqrt{2}$	0

is such that three dimensions are required to show the spatial relationships between the four objects. Unfortunately, with real data it is not usually known how many dimensions are needed for a representation. Hence a range of dimensions has to be tried.

The usefulness of multidimensional scaling comes from the fact that situations often arise where the relationship between objects is not known, but a distance matrix can be estimated. This is so, for example, in psychology where subjects can often say how similar or different individual pairs of objects are, but cannot draw an overall picture of the relationships between the objects. Multidimensional scaling can then provide a picture.

At the present time there are a wide variety of data analysis techniques that go under the general heading of multidimensional scaling. Here only the simplest of these will be considered, these being the classical methods proposed by Torgerson (1952) and Kruskal (1964a, b). A related method called principal coordinates analysis is discussed in chapter 12.

11.2 Procedure for multidimensional scaling

A classical multidimensional scaling starts with a matrix of distances between n objects which has δ_{ij}, the distance from object i to object j, in the ith row and jth column. The number of dimensions, t, for the mapping of objects is fixed for a particular solution. Different computer programs use different methods for carrying out analysis but generally something like the following steps are involved:

1. A starting configuration is set up for the n objects in t dimensions, i.e. coordinates $(x_1, x_2, \ldots, x_t)$ are assumed for each object in a t-dimensional space.
2. The Euclidean distances between the individuals are calculated for the configuration. Let d_{ij} be the distance between individual i and individual j.
3. A regression of d_{ij} on δ_{ij} is made where, as mentioned above, δ_{ij} is the distance between individual i and j according to the input data. The regression can be linear, polynomial or monotonic. For example, a linear regression assumes that

$$d_{ij} = a + b\delta_{ij} + e,$$

where e is an 'error' term and a and b are constants. A monotonic regression assumes simply that if δ_{ij} increases then d_{ij} increases or remains constant but no exact relationship between δ_{ij} and d_{ij} is fitted. The distances obtained from the regression equation $(\hat{d}_{ij} = a + b\delta_{ij}$, assuming a linear regression) are called 'disparities'. That is to say, the disparities $\hat{d}_{ij}$ are the data distances δ_{ij} scaled to match the configuration distance d_{ij} as closely as possible.

4. The goodness of fit between the configuration distances and the disparities is measured by a suitable statistic. One possibility is Kruskal's 'stress formula 1', which is

$$\text{STRESS } 1 = \{\sum (d_{ij} - \hat{d}_{ij})^2 / \sum \hat{d}_{ij}^2\}^{1/2} \tag{11.1}$$

The description 'stress' is used here because the statistic is a measure of the extent to which the spatial configuration of points has to be stressed in order to obtain the data distances δ_{ij}.

5. The coordinates $(x_1, x_2, \ldots, x_t)$ of each object are changed slightly in such a way that the stress is reduced.

Steps 2 to 5 are repeated until it seems that the stress cannot be further reduced. The outcome of the analysis is then the coordinates of the n individuals in t dimensions. These coordinates can be used to draw a 'map' which shows how the individuals are related. It is desirable that a good solution is found in three or fewer dimensions, as a graphical representation of the n objects is then straightforward. Obviously this is not always possible.

Obviously, small values of STRESS 1 (close to 0) are desirable. However, defining what is meant by 'small' for a good solution is not straightforward. As a rough guide, Kruskal and Wish (1978, p. 56) indicate that reducing the number of dimensions to the extent that STRESS 1 exceeds 0.1, or increasing the number of dimensions when STRESS 1 is already less than 0.05, is questionable. However, their discussion concerning choosing the number of dimensions encompasses more considerations than this.

An important distinction is between metric multidimensional scaling and non-metric dimensional scaling. In the metric case the configuration distances d_{ij} and the data distances δ_{ij} are related by a linear or polynomial regression equation. However, with non-metric scaling all that is required is a monotonic regression, which means that only the ordering of the data distances is important. Generally the greater flexibility of non-metric scaling should enable better low-dimensional representations of the data to be obtained.

Example 11.1 Road distances between New Zealand towns

As an example of what can be achieved by multidimensional scaling, a 'map' of the south island of New Zealand has been constructed from a table of the road distances between the 13 towns shown in Fig. 11.3.

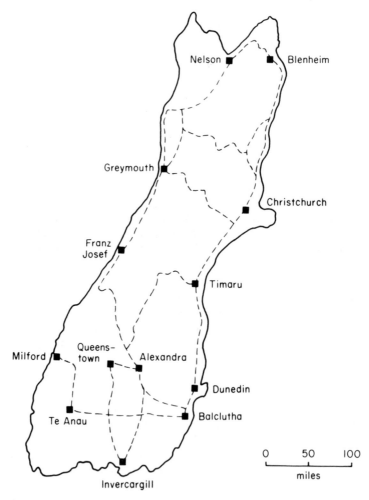

Figure 11.3 The south island of New Zealand. Main roads are indicated by broken lines. The 13 towns used for Example 11.1 are indicated.

Table 11.1 Main road distances (miles) between 13 towns in the south island of New Zealand. The positions of the towns and road links are shown in Fig. 11.3

	Alexandra	Balclutha	Blenheim	Christ-church	Dunedin	Franz Josef	Grey-mouth	Inver-cargill	Milford	Nelson	Queens-town	Te Anau	Timaru
Alexandra	–												
Balclutha	100	–											
Blenheim	485	478	–										
Christchurch	284	276	201	–									
Dunedin	126	50	427	226	–								
Franz Josef	233	493	327	247	354	–							
Greymouth	347	402	214	158	352	114	–						
Invercargill	138	89	567	365	139	380	493	–					
Milford	248	213	691	489	263	416	555	174	–				
Nelson	563	537	73	267	493	300	187	632	756	–			
Queenstown	56	156	494	305	192	228	341	118	178	572	–		
Te Anau	173	138	615	414	188	366	480	99	75	681	117	–	
Timaru	197	177	300	99	127	313	225	266	377	366	230	315	–

If road distances were proportional to geographic distances it would be possible to recover the true map exactly by a two-dimensional analysis. However, due to the absence of direct road links between many towns, road distances are in some cases far greater than geographic distances. Consequently, all that can be hoped for is a rather approximate recovery.

The computer program ALSCAL-4 (Young and Lewyckyj, 1979) was used for the analysis. At step 3 of the procedure described above a monotonic regression relationship was assumed between the map distances d_{ij} and the distances δ_{ij} given in Table 11.1 to give what is sometimes called classical non-metric multidimensional scaling.

The program produced a two-dimensional solution for the data in four iterations of steps 2 to 5 of the algorithm described above. The final stress value was 0.052 as calculated using equation (11.1).

The output from the program includes the coordinates of the 13 towns on the 'map' produced in the analysis. These are shown in Table 11.2. A plot of the towns using these coordinates is shown in Fig. 11.4. A comparison of this figure with Fig. 11.3 indicates that

Table 11.2 Coordinates produced by multidimensional scaling applied to the distances between 13 towns shown in Table 11.1. These are the coordinates that the towns are plotted against in Fig. 11.4

| | Dimension | |
Town	1	2
Alexandra	0.72	−0.32
Balclutha	0.84	0.78
Blenheim	−1.99	0.43
Christchurch	−0.92	0.34
Dunedin	0.52	0.46
Franz Josef	−0.69	−1.23
Greymouth	−1.32	−0.57
Invercargill	1.28	0.39
Milford	1.83	−0.33
Nelson	−2.33	0.07
Queenstown	0.81	−0.49
Te Anau	1.47	−0.26
Timaru	−0.19	0.64

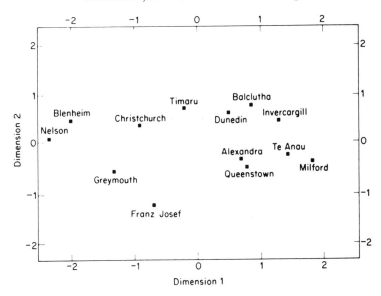

Figure 11.4 The map produced by a multidimensional scaling analysis of the data in Table 11.1.

the multidimensional scaling has been quite successful in recovering the map of South Island. On the whole the towns are shown with the correct relationships to each other. An exception is Milford. Because this can only be reached by road through Te Anau, the 'map' produced by multidimensional scaling has made Milford closest to Te Anau. In fact, Milford is geographically closer to Queenstown than it is to Te Anau.

All that is important with the configuration produced by multidimensional scaling is the relative positions of the objects being considered. This is unchanged by a rotation or a reflection. It is also unchanged by a magnification or contraction of all the scales. That is, the size of the configuration is not important. For this reason ALSCAL-4 always scales the configuration so that the average coordinate is zero in all dimensions and the sum of the squared coordinates is equal to the number of objects multiplied by the number of dimensions. Thus in Table 11.2 the sum of the coordinates is zero for each of the two dimensions and the total of the coordinates squared is 26.

Example 11.2 Voting behaviour of Congressmen

For a second example of the value of multidimensional scaling, consider the distance matrix shown in Table 11.3. Here the 'distances' are between 15 New Jersey Congressmen in the United States House of Representatives. They are simply a count of the number of voting disagreements on 19 bills concerned with environmental matters. For example, Congressmen Hunt and Sandman disagreed 8 out of the 19 times. Sandman and Howard disagreed 17 out of the 19 times, etc. An agreement was considered to occur if two Congressmen both voted yes, both voted no, or both failed to vote. The table of distances was constructed from original data given by Romesburg (1984, p. 155).

Two analyses were carried out using the ALSCAL-4 program. The first was a classical metric multidimensional scaling which assumes that the distances of Table 11.3 are measured on a ratio scale. That is to say, it is assumed that doubling a distance value is equivalent to assuming that the configuration distance between two objects is doubled. This means that the regression at step 3 of the procedure described above is of the form

$$d_{ij} = b\delta_{ij} + e,$$

where e is an error and b is some constant. The stress values obtained for 4-, 3- and 2-dimensional solutions were found on this basis to be 0.080, 0.121 and 0.194, respectively.

A second analysis was carried out by classical non-metric scaling so that the regression of d_{ij} on δ_{ij} was assumed to be monotonic only. In this case the stress values for 4-, 3- and 2-dimensional solutions were found to be 0.065, 0.089 and 0.134, respectively. The distinctly lower stress values for non-metric scaling suggest that this is preferable to metric scaling for these data and indeed the three-dimensional non-metric solution has only slightly more stress than the four-dimensional solution. This three-dimensional solution is therefore the one that will be considered in more detail.

Table 11.4 shows the coordinates of the Congressmen for the three-dimensional solution. A plot for the first two dimensions is shown in Fig. 11.5. The value for the third dimension is shown for each plotted point in the figure, where this dimension indicates how far a three-dimensional plot would make the point above or below the two-dimensional plane. For example, Daniels should be plotted 0.52

Table 11.3 The 'distances' between 15 Congressmen from New Jersey in the United States House of Representatives. The numbers in the table show the number of times that the Congressmen voted differently on 19 environmental bills. Party allegiances are indicated (R = Republican, D = Democrat)

	1	2	3	4	5	6	7	8	9	10	11	12	13	14	15
1 Hunt (R)	–														
2 Sandman (R)	8	–													
3 Howard (D)	15	17	–												
4 Thompson (D)	15	12	9	–											
5 Frelinghuysen (R)	10	13	16	14	–										
6 Forsythe (R)	9	13	12	12	8	–									
7 Widnall (R)	7	12	15	13	9	7	–								
8 Roe (D)	15	16	5	10	13	12	17	–							
9 Helstoski (D)	16	17	5	8	14	11	16	4	–						
10 Rodino (D)	14	15	6	8	12	10	15	5	3	–					
11 Minish (D)	15	16	5	8	12	9	14	5	2	1	–				
12 Rinaldo (R)	16	17	4	6	12	10	15	3	1	2	1	–			
13 Maraziti (R)	7	13	11	15	10	6	10	12	13	11	12	12	–		
14 Daniels (D)	11	12	10	10	11	6	11	7	7	4	5	6	9	–	
15 Pattern (D)	13	16	7	7	11	10	13	6	5	6	5	4	13	9	–

Table 11.4 Coordinates of the 15 Congressmen obtained from a three-dimensional non-metric multidimensional scaling of the distance matrix given in Table 11.3

Congressmen	Dimension		
	1	2	3
Hunt	2.25	0.15	0.53
Sandman	1.74	2.06	0.64
Howard	−1.37	−0.01	0.84
Thompson	−0.85	1.42	−0.45
Frelinghuysen	1.47	−0.83	−1.23
Forsythe	0.81	−0.93	−0.43
Widnall	2.25	−0.28	−0.46
Roe	−1.40	−0.01	0.60
Helstoski	−1.50	0.22	−0.18
Rodino	−1.09	−0.19	0.10
Minish	−1.13	−0.21	−0.24
Rinaldo	−1.27	−0.18	−0.27
Maraziti	1.20	−1.20	0.97
Daniels	−0.12	−0.16	0.52
Patten	−0.99	0.14	−0.94

units above the plane and Rinaldo should be plotted 0.27 units below the plane.

From Fig. 11.5 it is clear that dimension 1 is largely reflecting party differences. The Democrats fall on the left-hand side of the figure and the Republicans, other than Rinaldo, on the right-hand side.

To interpret dimension 2 it is necessary to consider what it is about the voting of Sandman and Thompson, who have the highest two scores, that contrasts with Maraziti and Forsythe, who have the two lowest scores. This points to the number of abstentions from voting. Sandman abstained from nine votes and Thompson abstained from six votes. Individuals with low scores on dimension 2 voted all or most of the time.

Dimension 3 appears to have no simple or obvious interpretation. It must reflect certain aspects of differences in voting patterns. However, these will not be considered for the present example. It suffices to say that the analysis has produced a representation of the

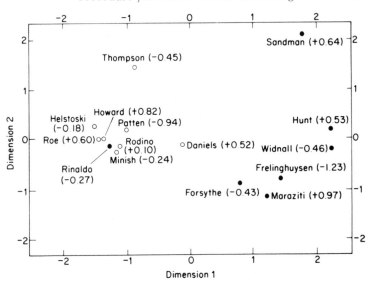

Figure 11.5 Plot of Congressmen against the first two dimensions of the configuration produced by a three-dimensional classical non-metric multidimensional scaling of the data in Table 11.3. Open circles indicate Democrats, closed circles indicate Republicans. The coordinate for dimension 3 is indicated in parentheses for each point.

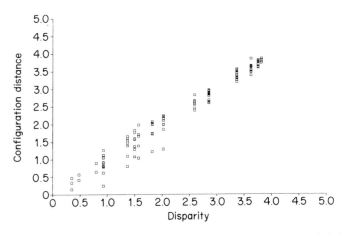

Figure 11.6 Plot of configuration distances against disparities (scaled data distances).

Congressmen in three dimensions that indicates how they relate with regard to voting on environmental issues.

Three graphs are helpful in assessing the accuracy of the solution that has been obtained. Figure 11.6 shows the first of these, which is a plot of the distances between points on the derived configuration, d_{ij}, against the disparities, $\hat{d}_{ij}$. The figure indicates the lack of fit of the solution because the disparities are the 105 data distances of Table 11.3 after they have been scaled to match the configuration as closely as possible. A plot like that of Fig. 11.6 would be a straight line if all of the distances and disparities were equal.

Figure 11.7 is a plot of the distances between the configuration points (d_{ij}) against the original data distances (δ_{ij}). This relationship does not have to be a straight line with non-metric scaling. However, scatter about an underlying trend line does indicate lack of fit of the model. For example in Table 11.3 there are eight distances of 5. Figure 11.7 shows that these correspond to configuration distances of between about 0.2 and 1.25. With an error-free solution, equal data distances will give equal configuration distances.

Finally, Fig. 11.8 shows the monotonic regression that has been estimated between the disparities $\hat{d}_{ij}$ and the data distances δ_{ij}. As the data distances increase, all that is required is that the disparities increase or remain constant. The figure shows, for example, that data distances of 4 or 5 both give estimated disparities of about 0.9.

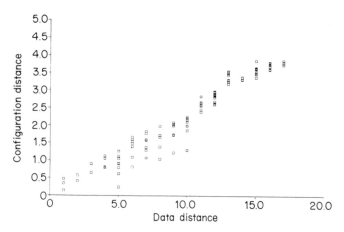

Figure 11.7 Plot of configuration distances against data distances.

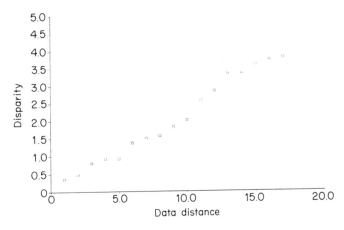

Figure 11.8 The monotonic regression of estimated disparities against data distances.

11.3 Computational methods and computer programs

The calculations for the examples in this chapter were carried out using ALSCAL-4 as a single program (Young and Lewyckyj, 1979). However, the same algorithm is also available as options in SAS (1985) and SPSS (1990). Other programs for multidimensional scaling are KYST (Kruskal *et al.*, 1977), MULTISCALE (Ramsay, 1978) and NMDS that is included with the book by Ludwig and Reynolds (1988). Schiffman *et al.* (1981) describe the use of ALSCAL, KYST, MULTISCALE and three other computer programs with several sets of data. Generally, different programs use different algorithms and therefore do not give exactly the same results. With good data it can be hoped that the differences will not be substantial.

11.4 Further reading

The book by Kruskal and Wish (1978) provides a short introduction to multidimensional scaling, but with more details than have been presented here. A more comprehensive treatment is provided by Schiffman *et al.* (1981).

Exercise

Consider the data on percentages employed in 26 countries in Europe in Table 1.5. From these data construct a 26 by 26 matrix of Euclidean distances between the countries using equation (5.1). Carry out nonmetric multidimensional scaling using this matrix to find out how many dimensions are needed to represent the countries in a manner that reflects differences between their employment patterns.

References

Kruskal, J.B. (1964a) Multidimensional scaling by optimizing goodness of fit to a nonmetric hypothesis. *Psychometrics* **29**, 1–27.

Kruskal, J.B. (1964b) Nonmetric multidimensional scaling: a numerical method. *Psychometrics* **29**, 115–29.

Kruskal, J.B. and Wish, M. (1978) *Multidimensional Scaling.* Sage Publications, Beverly Hills.

Kruskal, J.B., Young, F.W. and Seery, J.B. (1977) *How to Use KYST-2A, a Very Flexible Program to do Multidimensional Scaling.* Bell Laboratories, 600 Mountain Avenue, Murray Hill, New Jersey 07974.

Ludwig, J.A. and Reynolds, J.F. (1988) *Statistical Ecology.* Wiley, New York.

Ramsay, J.O. (1978) *MULTISCALE: Four Programs for Multidimensional Scaling by the Method of Maximum Likelihood.* National Educational Resources Inc., 1525 E. 53rd Street, Chicago, Illinois 60615.

Romesburg, H.C. (1984) *Cluster Analysis for Researchers.* Lifetime Learning Publications, Belmont, California.

Schiffman, S.S., Reynolds, M.L. and Young, F.W. (1981) *Introduction to Multi-Dimensional Scaling: Theory, Methods and Applications.* Academic Press, Orlando.

SAS (1985) *SAS User's Guide: Statistics.* SAS Institute, Cary, North Carolina 27511.

SPSS (1990) *SPSS Reference Guide.* SPSS Inc., 444 N. Michigan Avenue, Chicago, Illinois 60611.

Togerson, W.S. (1952) Multidimensional scaling. I. Theory and method. *Psychometrics* **17**, 401–19.

Young, F.W. and Lewyckyj, R. (1979) *ALSCAL-4 User's Guide.* Psychometric Laboratory, University of North Carolina, Chapel Hill, North Carolina 27514.

CHAPTER TWELVE

Ordination

12.1 The ordination problem

The word 'ordination' for a biologist means essentially the same as 'scaling' does to a social scientist. Both words describe the process of producing a small number of variables that can be used to describe the relationship between a group of objects starting either from a matrix of distances or similarities between the objects, or from the values of a large number of variables measured on each object. From this point of view, many of the methods that have been described in earlier chapters can be used for ordination, and some of the examples have been concerned with this process. In particular, plotting female sparrows against the first two principal components of size measurements (Example 5.1), plotting European countries against the first two principal components for employment variables (Example 5.2), producing a map of New Zealand from a table of distances between towns by multidimensional scaling (Example 11.1), and plotting New Jersey Congressmen against axes obtained by multidimensional scaling based on voting behaviour (Example 11.2) are all examples of ordination. In addition, discriminant function analysis can be thought of as a type of ordination that is designed to emphasize the differences between objects in different groups, while canonical correlation analysis can be thought of as a type of ordination that is designed to emphasize the relationship between objects based on the relationships between two groups of variables.

Although ordination can be considered to cover a diverse range of situations, in biology it is most often used as a means of summarizing the relationships between different species as determined from their abundances at a number of different locations or, alternatively, as a means of summarizing the relationships between different locations on the basis of the abundances of different species at those locations. It is this type of application that is considered in the present chapter, although the examples involve archaeology as well as biology. The purpose of the chapter is to give more examples of the use of principal components analysis and multidimensional scaling in this context, and to describe the methods of principal coordinates analysis and correspondence analysis that have not been covered in earlier chapters.

12.2 Principal components analysis

Principal components analysis has already been discussed in Chapter 6. It may be recalled that it is a method whereby the values for variables $X_1, X_2, \ldots, X_p$ measured on each of n objects are used to construct principal components $Z_1, Z_2, \ldots, Z_p$ that are linear combinations of the X variables and are such that Z_1 has the maximum possible variance, Z_2 has the largest possible variance conditional on it being uncorrelated with Z_1, Z_3 has the maximum possible variance conditional on it being uncorrelated with both Z_1 and Z_2, and so on. The idea is that it may be possible for some purposes to replace that X variables with a smaller number of principal components with little loss of information.

In terms of ordination, it can be hoped that the first two principal components are sufficient to describe the differences between the objects because then a plot of Z_2 against Z_1 with the objects identified provides what is required. It is less satisfactory to find that three principal components are important but a plot of Z_2 against Z_1 with values of Z_3 indicated may be acceptable. If four or more principal components are important then, of course, a good ordination is not obtained, at least as far as a graphical representation is concerned.

Example 12.1 Plant species in the Steneryd Nature Reserve

Table 9.3 shows the abundances of 25 plant species on 17 plots from a grazed meadow in Steneryd Nature Reserve in Sweden as described in Exercise 1 of Chapter 9, which was concerned with using the data for cluster analyses. Now it is an ordination of the plots that will be considered so that the variables for principal components analysis are the abundances of the plant species. In other words, in Table 9.3 the objects of interest are the plots (columns) and the variables are the species (rows).

Table 12.1 Eigenvalues from a principal components analysis of the data in Table 9.3 treating the plots as the objects of interest and the species counts as the variables

Component	Eigenvalue	% of total	Cumulative %
1	8.792	35.17	35.17
2	5.585	22.34	57.51
3	2.955	11.82	69.33
4	1.929	7.72	77.04
5	1.581	6.32	83.37
6	1.130	4.52	87.89
7	0.993	3.97	91.86
8	0.545	2.18	94.04
9	0.401	1.16	95.64
10	0.349	1.39	97.04
11	0.196	0.78	97.82
12	0.176	0.70	98.53
13	0.127	0.51	99.03
14	0.116	0.46	99.50
15	0.074	0.30	99.79
16	0.051	0.21	100.00

Because there are more species than plots, the number of non-zero eigenvalues in the correlation matrix is determined by the number of plots. In fact there are 16 non-zero eigenvalues, as shown in Table 12.1. The first three components account for about 69% of the variation in the data, which is not a particularly high amount. The coefficients for the first three principal components are shown in Table 12.2. They are all contrasts between the abundance of different species that may well be meaningful to a botanist, but no interpretations will be attempted here.

Figure 12.1 shows a draftsman's diagram of the plot number (1–17) and the first three principal components. It is noticeable that the first component is closely related to the plot number. This reflects the fact that the plots are in the order of the abundance in the plots of species with a high response to light and a low response to moisture, soil reaction and nitrogen. Hence the analysis has at least been able to detect this trend.

Table 12.2 The first three principal components for the data in Table 9.3. The values shown are the coefficients of the standardized species abundances (with zero means and unit standard deviations). Species names are obvious abbreviations of the full names

Species	Z_1	Z_2	Z_3
Fes-o	0.298	0.005	−0.065
Ane-n	−0.250	0.018	−0.185
Sta-h	−0.201	0.203	−0.193
Agr-t	0.173	0.243	0.007
Ran-f	−0.108	−0.316	−0.072
Mer-p	−0.081	−0.312	0.018
Poa-p	−0.114	0.315	−0.112
Rum-a	−0.007	0.336	0.227
Ver-c	−0.146	0.356	−0.057
Dac-g	−0.234	0.146	0.183
Fra-e	−0.255	−0.111	0.168
Sax-g	0.130	0.243	0.226
Des-f	−0.049	0.117	−0.445
Luz-c	0.281	0.093	0.003
Pla-l	0.266	0.105	0.255
Fes-r	−0.027	0.227	0.192
Hie-p	0.271	−0.021	0.048
Geu-u	−0.203	−0.179	0.287
Lat-m	−0.147	0.257	−0.190
Cam-p	−0.206	0.177	0.070
Vio-r	−0.235	0.165	0.108
Hep-n	−0.213	0.027	0.335
Ach-m	0.288	0.026	0.097
All-s	−0.180	−0.120	0.364
Tri-r	0.206	0.112	0.216

Example 12.2 Burials in Bannadi

For a second example of principal components ordination, the data shown in Table 9.4 concerning grave goods from a cemetery in Bannadi, northeast Thailand will be considered. The table (kindly supplied by Professor C.F.W. Higham) shows the presence or absence of 38 different types of article in each of 47 burials, with additional information on whether the body was of an adult male, adult female, or a child. In Exercise 2 of Chapter 9 it was suggested that cluster analysis be used to study the relationships between the burials. Now

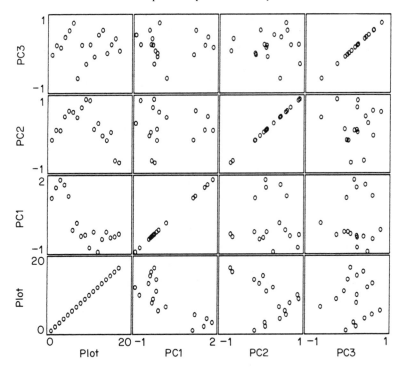

Figure 12.1 Draftsman's diagram for the ordination for 17 plots from Steneryd Nature Reserve. The four variables are the plot number and the first three principal components (PC1 to PC3).

ordination is considered with the same end in mind. For a principal components analysis the burials are the objects of interest and the 38 types of grave goods provide the 0–1 variables to be analysed. These variables were standardized before use so that the analysis was based on their correlation matrix.

In a situation like this where only presence and absence data are available it is common to find that a fairly large number of principal components are needed in order to account for most of the variation in the data. This is certainly the case here, with 11 components needed to account for 80% of the variance and 15 required to account for 90% of the variance. Obviously there are far too many important principal components for a satisfactory ordination.

For this example only the first four principal components will be

Table 12.3 Coefficients of standardized presence–absence variables for the first four principal components of the Bannadi data. Coefficients that are 0.2 or more (ignoring the sign) are underlined

Article	PC1	PC2	PC3	PC4
1	0.0	−0.0	0.0	−0.0
2	−0.1	−0.0	−0.0	0.5
3	−0.2	0.4	−0.0	−0.0
4	−0.1	−0.0	−0.0	0.5
5	−0.2	0.4	−0.0	−0.0
6	−0.2	0.4	−0.0	−0.0
7	−0.0	−0.0	−0.0	−0.0
8	−0.0	−0.0	−0.0	−0.0
9	0.2	0.1	0.3	0.1
10	0.2	0.1	0.0	0.1
11	0.0	0.0	0.0	−0.0
12	0.1	0.0	0.1	0.1
13	−0.0	−0.0	−0.0	−0.1
14	−0.2	0.4	−0.0	−0.0
15	0.0	−0.0	0.0	−0.0
16	0.2	0.1	0.3	0.1
17	−0.0	−0.0	−0.0	−0.1
18	0.2	0.2	−0.4	0.0
19	0.2	0.2	−0.4	0.0
20	0.2	0.2	−0.4	0.0
21	−0.1	−0.0	−0.0	0.5
22	−0.0	−0.0	0.0	0.0
23	0.3	0.2	−0.2	0.1
24	0.0	0.0	0.1	−0.1
25	0.3	0.2	0.3	0.1
26	0.3	0.2	0.3	0.1
27	0.1	0.0	0.0	0.0
28	−0.2	0.2	−0.0	0.3
29	−0.2	0.2	0.0	−0.1
30	0.2	0.1	0.0	0.0
31	0.1	0.0	0.2	−0.0
32	0.3	0.1	0.0	0.0
33	−0.0	−0.0	0.1	−0.1
34	0.2	0.1	−0.1	0.0
35	0.0	0.0	0.2	0.1
36	−0.1	0.2	0.2	−0.1
37	0.3	0.1	0.1	0.0
38	0.1	0.2	0.0	−0.0

considered, with the understanding that much of the variation in the original data is not accounted for. In fact, the four components correspond to eigenvalues of 5.29, 4.43, 3.65 and 3.34, while the total of all eigenvalues is 38 (the number of types of articles). Thus these components account for 13.9, 11.6, 9.6 and 8.8%, respectively, of the total variance, and between them account for 43.9% of the variance in the data.

The coefficients of the standardized presence–absence variables are shown in Table 12.3 with the largest values (arbitrarily set at an absolute value greater than 0.2) underlined. To aid in interpretation, the signs of the coefficients have been reversed if necessary from what was given by the computer output in order to ensure that the

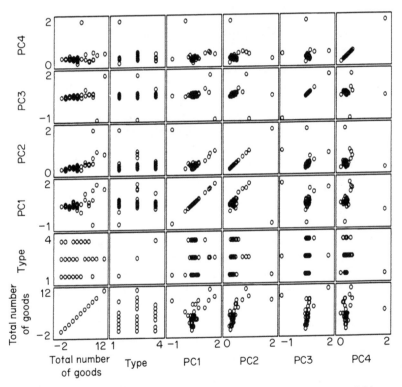

Figure 12.2 Draftsman's diagram for the 47 Bannadi graves. The six variables are the total number of different types of goods in the grave, an indicator of the type of body (1 = adult male; 2 = adult female; 3 = child) and the first four principal components (PC1 to PC4).

values of all the components are positive for burial B48 which has the largest number of items present. This is allowable because switching the signs of all the coefficients for a component does not change the percentage of variation explained by the component and, indeed, the direction of the signs is merely an accidental outcome of the numerical methods used to find eigenvectors of the correlation matrix.

From the large coefficients of component 1 it can be seen that this is indicating the presence of articles type 9, 10, 16, 18, 19, 20, 23, 25, 26, 30, 32, 34 and 37, and the absence of articles type 3, 5, 6, 14, 28 and 29. There is no grave with exactly this composition, but the component measures the extent to which each of the graves matches this model. The other components can also be interpreted in a similar way from the coefficients in Table 12.3.

Figure 12.2 shows a draftsman's plot of the total number of goods, the type of body, and the first four principal components. From studying this it is possible to draw some conclusions about the nature of the graves. For example, it seems that male graves tend to have low values and female graves to have high values for principal component 1, possibly reflecting a difference in grave goods associated with sex. Also, grave B47 has an unusual composition in comparison to the other graves. However, the fact that four principal components are being considered makes a simple interpretation of the results difficult.

12.3 Principal coordinates analysis

Principal coordinates analysis is similar to metric multidimensional scaling as discussed in Chapter 11. Both methods start with a matrix of similarities or distances between a number of objects and endeavour to find ordination axes. However, they differ in the numerical approach that is used. Principal coordinates analysis uses an eigen-value approach that can be thought of as a generalization of principal components analysis. However, multidimensional scaling, at least as defined in this book, attempts instead to minimize 'stress', where this is a measure of the extent to which the positions of objects in an m dimensional configuration fail to match the original distances or similarities after appropriate scaling.

To see the connection between principal coordinates analysis and principal components analysis it is necessary to recall some of the theoretical results concerning principal components analysis from Chapter 6, and to use some further results that are mentioned here for the first time. In particular:

1. The ith principal component is a linear combination

$$Z_i = a_{i1} X_1 + a_{i2} X_2 + \cdots + a_{ip} X_p$$

of the variables $X_1, X_2, \ldots, X_p$ that are measured on each of the objects being considered. There are p of these components, and the coefficients a_{ij} are given by the eigenvector $\mathbf{a}_i$ corresponding to the ith largest eigenvalue λ_i of the sample covariance matrix $\mathbf{C}$ of the X variables. That is to say, the equation

$$\mathbf{Ca}_i = \lambda_i \mathbf{a}_i. \tag{12.1}$$

is satisfied where $\mathbf{a}'_i = (a_{i1}, a_{i2}, \ldots, a_{ip})$. Also, the variance of Z_i is $\text{Var}(Z_i) = \lambda_i$, where this is zero if Z_i corresponds to a linear combination of the X variables that is constant.

2. If the X variables are coded to have zero means in the original data then the p by p covariance matrix $\mathbf{C}$ has the form

$$\mathbf{C} = \begin{bmatrix} \sum x_{i1}^2 & \sum x_{i1} x_{i2} & \cdots & \sum x_{i1} x_{ip} \\ \sum x_{i2} x_{i1} & \sum x_{i2}^2 & \cdots & \sum x_{i2} x_{ip} \\ \vdots & \vdots & & \vdots \\ \sum x_{ip} x_{i1} & \sum x_{ip} x_{i2} & \cdots & \sum x_{ip}^2 \end{bmatrix} \bigg/ (n-1)$$

where there are n objects, x_{ij} is the value of X_j for the ith object, and the summations are for i from 1 to n. Hence

$$\mathbf{C} = \mathbf{X}'\mathbf{X}/(n-1), \tag{12.2}$$

where

$$\mathbf{X} = \begin{bmatrix} x_{11} & x_{12} & \cdots & x_{1p} \\ x_{21} & x_{22} & \cdots & x_{2p} \\ \vdots & \vdots & & \vdots \\ x_{n1} & x_{n2} & \cdots & x_{np} \end{bmatrix}$$

is a matrix containing the original data values.

3. The symmetric n by n matrix

$$\mathbf{S} = \mathbf{XX'} = \begin{bmatrix} \sum x_{1j}^2 & \sum x_{1j}x_{2j} & \cdots & \sum x_{1j}x_{nj} \\ \sum x_{2j}x_{1j} & \sum x_{2j}^2 & \cdots & \sum x_{2j}x_{nj} \\ \vdots & \vdots & & \vdots \\ \sum x_{nj}x_{1j} & \sum x_{nj}x_{2j} & \cdots & \sum x_{nj}^2 \end{bmatrix}, \quad (12.3)$$

where the summations are for j from 1 to p, can be thought of as containing measures of the similarities between the n objects being considered. This is not immediately apparent, but is justified by considering the squared Euclidean distance from object i to object k, which is

$$d_{ik}^2 = \sum_{j=1}^{p} (x_{ij} - x_{kj})^2.$$

Expanding the right-hand side of this equation shows that

$$d_{ik}^2 = \sum_{j=1}^{p} x_{ij}^2 + \sum_{j=1}^{p} x_{kj}^2 - 2 \sum_{j=1}^{p} x_{ij}x_{kj}$$

or

$$d_{ik}^2 = s_{ii} + s_{kk} - 2s_{ik}, \quad (12.4)$$

where s_{ik} is the element in the ith row and kth column of $\mathbf{XX'}$. It follows that s_{ik} is a measure of the similarity between objects i and k because increasing s_{ik} means that the distance d_{ik} between the objects is decreased. Further, it is seen that s_{ik} takes the maximum value of $(s_{ii} + s_{kk})/2$ when $d_{ik} = 0$, which occurs when the objects i and k have identical values for the variables X_1 to X_p.

4. If the matrix

$$\mathbf{Z} = \begin{bmatrix} z_{11} & z_{12} & \cdots & z_{1p} \\ z_{21} & z_{22} & \cdots & z_{2p} \\ \vdots & \vdots & & \vdots \\ z_{n1} & z_{n2} & \cdots & z_{np} \end{bmatrix}$$

contains the values of the p principal components for the n objects being considered, then this can be written in terms of the data matrix $\mathbf{X}$ as

$$\mathbf{Z} = \mathbf{XA'}, \qquad (12.5)$$

where the ith row of $\mathbf{A}$ is $\mathbf{a}_i'$, the ith eigenvector of the sample covariance matrix $\mathbf{C}$. It is a property of $\mathbf{A}$ that $\mathbf{A'A} = \mathbf{I}$, i.e. the transpose of $\mathbf{A}$ is the inverse of $\mathbf{A}$. Thus postmultiplying both sides of equation (12.5) by $\mathbf{A}$ gives

$$\mathbf{X} = \mathbf{ZA}. \qquad (12.6)$$

This statement of results has been lengthy but it has been necessary in order to explain principal coordinates analysis in relationship to principal components analysis. To see this relationship, note that from equations (12.1) and (12.2)

$$\mathbf{X'Xa}_i/(n-1) = \lambda_i\mathbf{a}_i.$$

Then premultiplying both sides of this equation by $\mathbf{X}$ and using equation (12.3) gives

$$\mathbf{S}(\mathbf{Xa}_i) = (n-1)\lambda_i(\mathbf{Xa}_i)$$

or

$$\mathbf{Sz}_i = (n-1)\lambda_i\mathbf{z}_i, \qquad (12.7)$$

where $\mathbf{z}_i = \mathbf{Xa}_i$ is a vector of length n which contains the values of Z_i for the n objects being considered. Therefore, the ith largest eigenvalue of the similarity matrix $\mathbf{S} = \mathbf{X'X}$ is $(n-1)\lambda_i$ and the corresponding eigenvector gives the values of the ith principal component for the n objects.

Principal coordinates analysis consists of applying equation (12.7) to an n by n matrix $\mathbf{S}$ of similarities between n objects that is calculated using any of the many available similarity indices. In this way it is possible to find the 'principal components' corresponding to $\mathbf{S}$ without necessarily measuring any variables on the objects of interest. The components will have the properties of principal components and, in particular, will be uncorrelated for the n objects.

Applying principal coordinates analysis to the matrix $\mathbf{XX'}$ will give essentially the same ordination as a principal components analysis on the data in **X**. The only difference will be in terms of the scaling given to the components. In principal components analysis it is usual to scale the ith component to have the variance λ_i but with a principal coordinates analysis the component would usually be scaled to have a variance of $(n-1)\lambda_i$. This difference is immaterial because it is only the relative values of objects on ordination axes that is important.

There are two complications that can arise in a principal co-ordinates analysis that must be mentioned. They occur when the similarity matrix analysed does not have all the properties of a matrix calculated from data using the equation $\mathbf{S} = \mathbf{XX}'$.

First, from equation (12.3) it can be seen that the sums of the rows and columns of $\mathbf{XX}'$ are all zero. For example, the sum of the first row is

$$\sum x_{1j}^2 + \sum x_{1j} x_{2j} + \cdots + \sum x_{1j} x_{nj} = \sum x_{1j}(x_{1j} + x_{2j} + \cdots + x_{nj}),$$

where the summations are for j from 1 to p. This is zero because $x_{1j} + x_{2j} + \cdots + x_{nj}$ is n times the mean of X_j, and all X variables are assumed to have zero means. Hence it is required that the similarity matrix **S** should have zero sums for rows and columns. If this is not the case then the initial matrix can be double-centred by replacing the element s_{ik} in row i and column k by $s_{ik} - s_{i.} - s_{.k} + s_{..}$, where $s_{i.}$ is the mean of the ith row of **S**, $s_{.k}$ is the mean of the kth column of **S**, and $s_{..}$ is the mean of all the elements in **S**. The double-centred similarity matrix will have zero row and column means and is therefore more suitable for the analysis.

The second complication is that some of the eigenvalues of the similarity matrix may be negative. This is disturbing because the corresponding principal components appear to have negative vari-ances! However, the truth is just that the similarity matrix could not have been obtained by calculating $\mathbf{S} = \mathbf{XX}'$ for any data matrix. With ordination only the components associated with the largest eigenvalues are usually used so that a few small negative eigenvalues can be regarded as being unimportant. Large negative eigenvalues suggest that the similarity matrix being used is not suitable for ordination.

Computer programs for principal coordinates analysis sometimes offer the option of starting with either a distance matrix or a similarity matrix. If a distance matrix is used then it can be converted to a similarity matrix by transforming the distance d_{ik} to the similarity measure $s_{ik} = -d_{ik}^2/2$. The squared distance is used here because equation (12.4) indicates that this is appropriate.

Example 12.3 Plant species in the Steneryd Nature Reserve (Revisited)

As an example of the use of principal coordinates analysis the data considered in Example 12.1 on species abundances on plots in Steneryd Nature Reserve were reanalysed using Manhattan distances between plots. That is, the distance between plots i and k was measured by $d_{ik} = \sum |x_{ij} - x_{kj}|$, where the summation is for j over

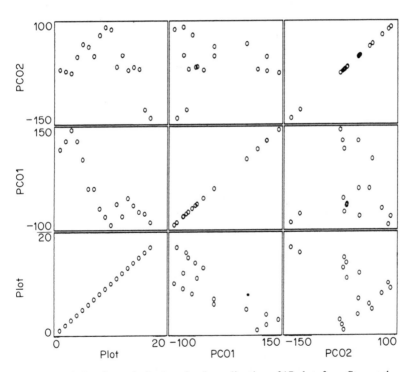

Figure 12.3 Draftsman's diagram for the ordination of 17 plots from Steneryd Nature Reserve based on a principal coordinates analysis on Manhattan distances between plots. The three variables are the plot number and the first two components (PCO1 and PCO2).

the 25 species and x_{ij} denotes the abundance of species j on plot i as given in Table 9.3. Similarities were calculated as $s_{ik} = -d_{ik}^2/2$ and then double-centred before eigenvalues and eigenvectors were found.

The first two eigenvalues of the similarity matrix were found to be 97 638.6 and 55 659.5, which account for 47.3 and 27.0% of the sum of the eigenvalues. On the face of it the first two components therefore give a good ordination with 74.3% of the variation accounted for. The third eigenvalue is much smaller at 12 488.2 and only accounts for 6.1% of the total.

Figure 12.3 shows a draftsman's diagram of the plot number and the first two components. Both components show a relationship with the plot number which, as noted in Example 12.1, is itself related to the response of the different species to environmental variables. Actually, a comparison of this draftsman's diagram with the six graphs in the bottom left-hand corner of Fig. 12.1 shows that the first two axes from the principal coordinates analysis are really very similar to the first two principal components apart from a difference in scaling.

Example 12.4 Burials in Bannadi (Revisited)

As an example of a principal coordinates analysis on presence and absence data, consider again the data in Table 9.4 on grave goods in the Bannadi cemetery in northeast Thailand. The analysis started with the matrix of unstandardized Euclidean distances between the 47 burials so that the distance from grave i to grave k was taken to be $d_{ik} = \sqrt{\{\sum(x_{ij} - x_{kj})^2\}}$, where the summation is for j from 1 to 38, and x_{ij} is 1 if the jth type of article is present in the ith burial, or otherwise is 0. A similarity matrix was then obtained as described in Example 12.3 and double-centred before eigenvalues and eigenvectors were obtained.

The principal coordinates analysis carried out in this manner gives the same result as a principal components analysis using unstandardized values for the X variables (i.e. carrying out a principal components analysis using the sample covariance matrix instead of the sample correlation matrix). The only difference in the results is in the scalings that are usually given to the ordination variables by principal components analysis and principal coordinates analysis.

The first four eigenvalues of the similarity matrix were 24.9, 19.3,

10.0 and 8.8, corresponding to 21.5, 16.6, 8.7 and 7.6%, respectively, of the sum of all eigenvalues. These components account for a mere 54.5% of the total variation in the data, but this is better than the 43.9% accounted for by the first four principal components obtained from the standardized data (Example 12.2).

Figure 12.4 shows a draftsman's diagram for the total number of goods in the burials, the type of body (adult male, adult female, or child), and the first four components. The signs of the first and fourth components were switched from those shown on the computer output so as to make them have positive values for burial B48 which contained the largest number of different types of grave goods. It

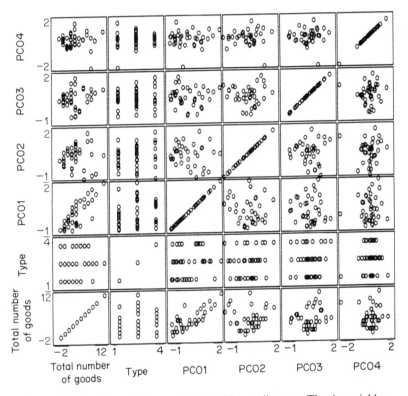

Figure 12.4 Draftsman's diagram for the 47 Bannadi graves. The six variables are the total number of different types of goods in the grave, an indicator of the type of body (1 = adult male; 2 = adult female; 3 = child), and the first four components from a principal coordinates analysis (PCO1 to PCO4).

can be seen from the diagram that the first component represents total abundance quite closely, but the other components are not related to this variable. Apart from this, the only obvious thing to notice is that one of the burials had a very low value for the fourth component. This is burial B47, which contained eight different types of article, of which four types were not seen in any other burial.

12.4 Multidimensional scaling

Multidimensional scaling has been discussed already in Chapter 11, where this is defined to be an iterative process for finding coordinates for objects on axes, in a specified number of dimensions, such that the distances between the objects match as closely as possible the distances or similarities that are provided in an input data matrix (section 11.2). The method will not be discussed further in the present chapter except as required to present the results of using it on the two example sets of data that have been considered with the other methods of ordination.

Example 12.5 Plant species in the Steneryd Nature
Reserve (again)

A multidimensional scaling of the 17 plots for the data in Table 9.3 was carried out using the computer program NMDS provided by Ludwig and Reynolds (1988). This performs a classical non-metric type of analysis on a distance matrix, so that the relationship between the data distances and the ordination (configuration) distances is assumed to be only monotonic. A feature of the program is that after a solution is obtained the axes are transformed to principal components. This ensures that the first axis accounts for the maximum possible variance in the ordination scores, the second axis accounts for the maximum possible remaining variance, and so on. The scores for the different axes are also made uncorrelated by this process.

For the example being considered, unstandardized Euclidean distances between the plots were used as input to the program. The stress values corresponding to solutions in from one to four dimensions were found to be 0.436, 0.081, 0.060, 0.023 and 0.021, so that a four-dimensional solution seems quite reasonable. Figure 12.5 shows a draftsman's diagram of the values for the plot numbers and the positions on these axes after they have been transformed to

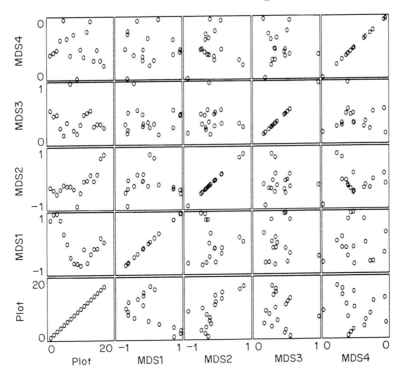

Figure 12.5 Draftsman's diagram for the ordination of 17 plots from Steneryd Nature Reserve based on non-metric multidimensional scaling on Euclidean distances between plots. The variables are the plot number and coordinates on four axes (MDS1 to MDS4).

principal components. Comparison with Figs 12.1 and 12.3 shows that the first multidimensional scaling axis corresponds closely with the first principal component and the first principal coordinate axis, while the second multidimensional scaling axis, after a reversal in sign, corresponds closely with the second principal component and the second principal coordinate axis.

Example 12.6 Burials in Bannadi (again)

The same analysis as used in the last example was also applied to the data on burials at Bannadi shown in Table 9.4. Unstandardized Euclidean distances between the 47 burials were calculated using the 0–1 data in the table as values for 38 variables, and these distances provided the data for Ludwig and Reynolds's (1988) computer program NMDS. The stress levels obtained for solutions in one to five dimensions were 0.405, 0.221, 0.113, 0.084 and 0.060. Hence the three-dimensional solution seems reasonable although the stress of 0.113 is quite large.

A draftsman's diagram for the three-dimensional solution is shown in Fig. 12.6, with the axes reversed where necessary to ensure that a positive value is obtained for burial B48 that has the highest number of different types of goods. Comparison with Fig. 12.2 shows that the first axis has a strong resemblance to the first principal

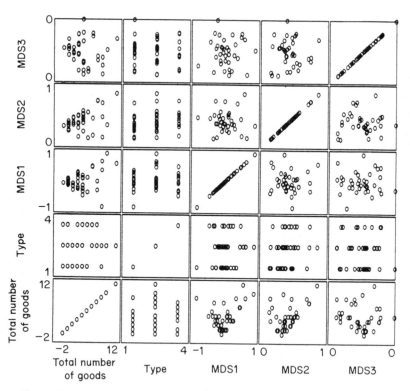

Figure 12.6 Draftsman's diagram for the 47 Bannadi graves. The variables are the total number of different types of goods in the grave, an indicator of the type of body (1 = adult male; 2 = adult female; 3 = child), and three axes (MSD1 to MSD3) from non-metric multidimensional scaling using unstandardized Euclidean distances between the graves.

component but otherwise the relationship with ordinations from other methods is not immediately clear.

12.5 Correspondence analysis

Correspondence analysis as a method of ordination originated in the work of Hirschfeld (1935), Fisher (1940) and a school of French statisticians (Benzecri, 1973). It is now the most popular method of ordination for plant ecologists and is receiving increasing use in other areas as well.

The method will be explained here in the context of the ordination of n sites on the basis of the abundance of p species, although it can be used equally well on data that can be presented as a two-way table of measures of abundance with the rows corresponding to one type of classification and the columns to a second type of classification.

With sites and species the situation is as shown in Table 12.4. Here there are a set of species values $a_1, a_2, \ldots, a_n$ associated with the rows of the table, and a set of site values $b_1, b_2, \ldots, b_p$ associated with the columns of the table. One interpretation of correspondence analysis is then that it is concerned with choosing species and site values so that they are as highly correlated as possible for the bivariate distribution that is represented by the abundances in the body of the table. That is to say, the site and species values are chosen to maximize their correlation for the distribution where the number of times that species i occurs at site j is proportional to the observed abundance x_{ij}.

It turns out that the solution to this maximization problem is

Table 12.4 The abundance of n species at p sites with species values and site values

Species	Site 1	2	$\cdots$	p	Row sum	Species value
1	x_{11}	x_{12}	$\cdots$	x_{1p}	R_1	a_i
2	x_{21}	x_{22}	$\cdots$	x_{2p}	R_2	a_2
$\vdots$					$\vdots$	$\vdots$
n	x_{n1}	x_{n2}	$\cdots$	x_{np}	R_n	a_n
Column sum	C_1	C_2	$\cdots$	C_p		
Site values	b_1	b_2	$\cdots$	b_p		

given by the set of equations

$$a_1 = \{(x_{11}/R_1)b_1 + (x_{12}/R_1)b_2 + \cdots + (x_{1p}/R_1)b_p\}/r$$
$$a_2 = \{(x_{21}/R_2)b_1 + (x_{22}/R_2)b_2 + \cdots + (x_{2p}/R_2)b_p\}/r$$
$$\vdots$$
$$a_n = \{(x_{n1}/R_n)b_1 + (x_{n2}/R_n)b_2 + \cdots + (x_{np}/R_n)b_p\}/r$$

and

$$b_1 = \{(x_{11}/C_1)a_1 + (x_{21}/C_1)a_2 + \cdots + (x_{n1}/C_1)a_n\}/r$$
$$b_2 = \{(x_{12}/C_2)a_1 + (x_{22}/C_2)a_2 + \cdots + (x_{n2}/C_2)a_n\}/r$$
$$\vdots$$
$$b_p = \{(x_{1p}/C_p)a_1 + (x_{2p}/C_p)a_2 + \cdots + (x_{np}/C_p)a_n\}/r$$

where R_i denotes the total abundance of species i, C_j denotes the total abundance at site j and r is the maximum correlation being sought. Thus the ith species value a_i is a weighted average of the site values, with site j having a weight that is proportional to x_{ij}/r_i, and the jth site value b_j is a weighted average of the species values, with species i having a weight that is proportional to x_{ji}/C_j.

The name *reciprocal averaging* is sometimes used to describe the equations just stated because the species values are (weighted) averages of the site values, and the site values are (weighted) averages of the species values. These equations are themselves often used as the starting point for justifying correspondence analysis as a means of producing species values as a function of site values and vice versa. It turns out that the equations can be solved iteratively after they have been modified to remove the trivial solution with $a_i = 1$ for all i, $b_j = 1$ for all j, and $r = 1$. However, it is more instructive to write the equations in matrix form in order to solve them because that shows that there may be several possible solutions to the equations and that these can be found from an eigenvalue analysis.

In matrix form, the equations shown above become

$$\mathbf{a} = \mathbf{R}^{-1}\mathbf{X}\mathbf{b}/r \tag{12.8}$$

and

$$\mathbf{b} = \mathbf{C}^{-1}\mathbf{X}'\mathbf{a}/r, \tag{12.9}$$

where $\mathbf{a}' = (a_1, a_2, \ldots, a_n)$, $\mathbf{b}' = (b_1, b_2, \ldots, b_p)$, $\mathbf{R}$ is an n by n diagonal matrix with R_i in the ith row and ith column, C is a p by p diagonal matrix with C_j in the jth row and jth column, and X is an n by p matrix with x_{ij} in the ith row and jth column. If equation (12.9) is substituted into equation (12.8) then after some matrix algebra it is found that

$$r^2(R^{1/2}a) = (R^{-1/2}XC^{-1/2})(R^{-1/2}XC^{-1/2})'(R^{1/2}a), \quad (12.10)$$

where $R^{1/2}$ is a diagonal matrix with $\sqrt{R_i}$ in the ith row and ith column, and $C^{1/2}$ is a diagonal matrix with $\sqrt{C_j}$ in the jth row and jth column. This shows that the solutions to the problem of maximizing the correlation are given by the eigenvalues of the n by n matrix $(R^{-1/2}XC^{-1/2})(R^{-1/2}XC^{-1/2})'$: for any eigenvalue λ_k the correlation between the species and site scores will be $r_k = \sqrt{\lambda_k}$; the eigenvector for this correlation will be $R^{1/2}a_k = (\sqrt{R_1}a_{1k}, \sqrt{R_2}a_{2k}, \ldots, \sqrt{R_n}a_{nk})'$, where a_{ik} are the species values; the corresponding site values can be obtained from equation (12.9) as $b_k = C^{-1}X'a_k/r_k$.

The largest eigenvalue will always be $r^2 = 1$, giving the trivial solution $a_i = 1$ for all i and $b_j = 1$ for all j. The remaining eigenvalues will be positive or zero and reflect different possible dimensions for representing the relationships between species and sites. These dimensions can be shown to be orthogonal in the sense that the species and site values for one dimension will be uncorrelated with the species and site values in other dimensions for the data distribution of abundances x_{ij}.

Ordination by correspondence analysis involves using the species and site values for the first few largest eigenvalues that are less than 1 because these are the solutions for which the correlations between species values and site values are strongest. It is common to plot both the species and the sites on the same axes because, as noted earlier, the species values are an average of the site values and vice versa. In other words, correspondence analysis gives an ordination of both species and sites at the same time.

It is apparent from equation (12.10) that correspondence analysis cannot be used on data which include a zero row sum because then the diagonal matrix $R^{-1/2}$ will have an infinite element. By a similar argument, zero column sums are not allowed either. This means that the method cannot be used on the burial data in Table 9.4 because some graves did not contain any articles. However, correspondence analysis can be used with presence and absence data when this problem is not present.

Example 12.7 Plant species in the Steneryd Nature Reserve
(yet again)

Correspondence analysis was applied to the data for species abundances in the Steneryd Nature Reserve (Table 9.3). There were 16
eigenvalues less than 1 and the values are as follows, with their square
roots (the correlations between species values and plot values) in
parentheses: 0.665 (0.82), 0.406 (0.64), 0.199 (0.45), 0.136 (0.37), 0.094
(0.31), 0.074 (0.27), 0.057 (0.24), 0.028 (0.17), 0.020 (0.14), 0.019 (0.14),
0.010 (0.10), 0.008 (0.09), 0.007 (0.08), 0.005 (0.07), 0.003 (0.05), 0.001
(0.03). The first two or three might be considered to be important,
but here only the species and plot values for the first two eigenvalues
will be used for ordination.

Figure 12.7 shows a graph of the species and plot values for the
eigenvalue of 0.406 (CORR2) against the species and plot values for
the eigenvalue of 0.665 (CORR1). Abbreviated names are shown for the
species, and S1 to S17 indicate the sites. The ordination of sites is

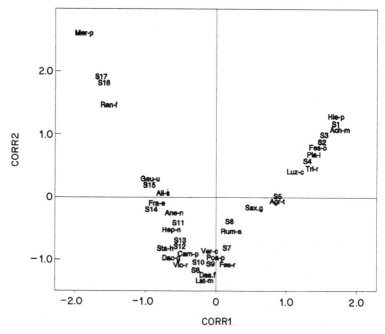

Figure 12.7 Plot of species and sites against the first two axes (CORR1 and
CORR2) found by applying correspondence analysis to the data from
Steneryd Nature Reserve. The species names have been given obvious
abbreviations and the sites are labelled S1 to S17.

quite clear, with an almost perfect sequence from S1 on the right to S17 on the left, moving round the very distinct arch. The species are interspersed among the sites along the same arch from Mer-p (*Mercurialis perennis*) on the left to Hie-p (*Hieracium pilosella*) on the right. A comparison of the figure with Table 12.1 shows that it makes a good deal of sense. For example, *M. perennis* is only abundant on the highest numbered sites and *H. pilosella* is only abundant on the lowest numbered sites.

The arch or 'horseshoe' that appears in the ordination for this example is a common feature in the results of correspondence analysis that is also sometimes apparent with other methods of ordination as well. There is sometimes concern that this effect will obscure the nature of the ordination axes and therefore some attention has been devoted to the development of ways to modify analyses to remove the effect which is considered to be an artefact of the ordination method. With correspondence analysis a method of detrending is usually used, and the resulting ordination method is then called detrended correspondence analysis (Hill and Gauch, 1980). Adjustments for other methods of ordination exist as well, but seem to receive little use.

12.6 Comparison of ordination methods

Four methods of ordination have been reviewed in this chapter and it is desirable to be able to state when each should be used. Unfortunately this cannot be done in an altogether satisfactory way because of the wide variety of different circumstances for which ordination is used. Therefore all that will be done here is to make some final comments on each of the methods in terms of its utility.

Principal components analysis can only be used when data on p variables is available for each of the objects being studied. Therefore it is not available in situations where only a distance or similarity matrix is available. Where variable values are available and the variables are approximately normally distributed this method is the obvious one to use.

When an ordination is required starting with a matrix of distance or similarities between the objects being studied it is possible to use either principal coordinates analysis or multidimensional scaling. Multidimensional scaling can be metric or non-metric, and principal coordinates analysis and metric multidimensional scaling should give similar results. The relative advantages of metric and non-metric multidimensional scaling will depend very much on the circumstances but, in general, non-metric scaling can be expected to give a slightly better fit to the data matrix.

Correspondence analysis was developed for situations where the objects of interest are being described by measures of the abundance of different characteristics. When this is the case this method appears to give ordinations that are relatively easy to interpret. It has certainly found favour with ecologists analysing data on the abundance of different species at different locations.

12.7 Computational methods and computer programs

Programs for principal components analysis have been discussed in Chapter 7 and need not be considered further here. The program MVSP (Kovach, 1990) was used for the calculations with Examples 12.1 and 12.2. This program was also used for the principal coordinates analyses of Examples 12.3 and 12.4, including the construction of distance matrices, and for the correspondence analysis of Example 12.7. It includes a considerable number of options for ordination, including detrended correspondence analysis.

Other programs for ordination include DECORANA for detrended correspondence analysis (Hill, 1979), CANOCO (ter Braak, 1988), those in the book by Ludwig and Reynolds (1988) and PATN (Belbin, 1989). In addition, standard statistical packages often include some ordination options as well as principal components analysis.

12.8 Further reading

Suggestions for further reading related to principal components analysis and multidimensional scaling are provided in Chapters 6 and 11 and it is unnecessary to repeat these here. For further discussions and more examples of principal coordinates analysis and correspondence analysis in the context of plant ecology see the books by Digby and Kempton (1987) and Ludwig and Reynolds (1988). For correspondence analysis the standard reference in English is by Greenacre (1984).

Exercise

Table 6.6 shows the values for six measurements taken on each of 25 prehistoric goblets excavated in Thailand. The nature of the measurements is shown in Fig. 6.3. Use the various methods discussed in this chapter to produce ordinations of the goblets and see which method appears to produce the most useful result.

References

Belbin, L. (1989) *PATN, Pattern Analysis Package*. Division of Wildlife and Ecology, CSIRO, Lyneham, A.C.T. 2602, Australia.

Benzecri, P.J. (1973) L'analyse des correspondances. *L'Analyse des Données*, Vol. 2. Dunod, Paris.

Digby, P.G.N. and Kempton, R.A. (1987) *Multivariate Analysis of Ecological Communities*. Chapman and Hall, London.

Fisher, R.A. (1940) The precision of discriminant functions. *Annals of Eugenics* **10**, 422–9.

Greenacre, M.J. (1984) *Theory and Application of Correspondence Analysis*. Academic Press, London.

Hill, M.O. (1979) *DECORANA – a FORTRAN Program for Detrended Correspondence Analysis and Reciprocal Averaging*. Section of Ecology and Systematics, Cornell University, Ithaca, New York.

Hill, M.O. and Gauch, H.G. (1980) Detrended correspondence analysis, an improved ordination technique. *Vegetatio* **42**, 47–58.

Hirschfeld, H.O. (1935) A connection between correlation and contingency. *Proceedings of the Cambridge Philosophical Society* **31**, 520–4.

Kovach, W.L. (1990) *MVSP, Multi-Variate Statistical Package, Plus Version 2.0*. Kovach Computing Services, 85 Nant-y-Felin, Pentraeth, Anglesey LL75 8UY, Wales.

Ludwig, J.A. and Reynolds, J.F. (1988) *Statistical Ecology*. Wiley, New York.

ter Braak, C.J.F. (1988) *CANOCO – a FORTRAN Program for Canonical Community Ordination by (Partial) (Detrended) (Canonical) Correspondence Analysis, Principal Components Analysis and Redundancy Analysis, Ver. 2.1*. Rep. LWA-88-02, Agricultural Mathematics Group, Wageningen, The Netherlands.

CHAPTER THIRTEEN

Epilogue

13.1 The next step

In writing this book my aims have purposely been rather limited. These aims will have been achieved if someone who has read the previous chapters carefully has a fair idea of what can and what cannot be achieved by the multivariate statistical methods that are most widely used. My hope is that the book will help many people take the first step in 'a journey of a thousand miles'.

For those who have taken this first step, the way to proceed further is to gain experience of multivariate methods by analysing different sets of data and seeing what results are obtained. Like other areas of applied statistics, competence in multivariate analysis requires practice.

13.2 Some general reminders

In developing expertise and familiarity with multivariate analyses there are a few general points that are worth keeping in mind. Actually, these points are just as relevant to univariate analyses. However, they are still worth emphasizing in the multivariate context.

First, it should be remembered that there are often alternative ways of approaching the analysis of a particular set of data, none of which is necessarily the 'best'. Indeed, several types of analysis may well be carried out to investigate different aspects of the same data. For example, the body measurements of female sparrows given in Table 1.1 can be analysed by principal components analysis or factor analysis to investigate the dimensions behind body size variation, by discriminant analysis to contrast survivors and non-survivors, by cluster analysis or multidimensional scaling to see how the birds group together, and so on.

Second, use common sense. Before embarking on an analysis consider whether it can possibly answer the questions of interest.

How many statistical analyses are carried out because the data are of the right form, irrespective of what light the analyses can throw on a question? At some time or another most users of statistics find themselves sitting in front of a large pile of computer output with the realization that it tells them nothing that they really want to know.

Third, multivariate analysis does not always work in terms of producing a 'neat' answer. There is an obvious bias in statistical textbooks and articles towards examples where results are straightforward and conclusions are clear. In real life this does not happen quite so often. Do not be surprised if multivariate analyses fail to give satisfactory results on the data that you are really interested in! It may well be that the data have a message to give, but the message cannot be read using the somewhat simple models that standard analyses are based on. For example, it may be that variation in a multivariate set of data can be completely described by two or three underlying factors. However, these may not show up in a principal components analysis or a factor analysis because the relationship between the observed variables and the factors is not a simple linear one.

Finally, there is always the possibility that an analysis is dominated by one or two rather extreme observations. These 'outliers' can sometimes be found by simply scanning the data by eye, or by considering frequency tables for the distributions of individual variables. In some cases a more sophisticated multivariate method may be required. A large Mahalanobis distance from an observation to the mean of all observations is one indication of a multivariate outlier (see section 5.3).

It may be difficult to decide what to do about an outlier. If it is due to a recording error or some other definite mistake then it is fair enough to exclude it from the analysis. However, if the observation is a genuine value then this is not valid. Appropriate action then depends on the particular circumstances. Barnett and Lewis (1984) and Hawkins (1980) have considered the problem of outliers at some length.

13.3 Missing values

Missing values can cause more problems with multivariate data than with univariate data. The trouble is that when there are many variables being measured on each individual it is often the case that one or two of these variables have missing values. It may then happen that if individuals with any missing values are excluded from an analysis this means excluding quite a large proportion of individuals, which may be completely impractical. For example, in studying ancient human populations skeletons are frequently broken and incomplete.

Texts on multivariate analysis are often remarkably silent on the question of missing values. To some extent this is because doing something about missing values is by no means a straightforward matter. In practice, computer packages sometimes include a facility for estimating missing values. For example, the BMDP package (Dixon, 1990) allows missing values to be estimated by several different 'common sense' methods. One possible approach is therefore to estimate missing values and then analyse the data, including these estimates, as if they were complete data in the first place. It seems reasonable to suppose that this procedure will work satisfactorily providing that only a small proportion of values are missing.

For more information about methods for dealing with missing data see Seber (1984) and Little and Rubin (1987).

References

Barnett, V. and Lewis, T. (1984) *Outliers in Statistical Data*. Wiley, New York.
Dixon, W.J. (ed.) (1990) *BMDP Statistical Software Manual*. University of California Press, Berkeley.
Hawkins, D.M. (1980) *Identification of Outliers*. Chapman and Hall, London.
Little, R.A. and Rubin, D.B. (1987) *Statistical Analysis with Missing Data*. Wiley, New York.
Seber, G.A.F. (1984) *Multivariate Observations*. Wiley, New York.

Author index

Andrews, D.F.   33, 34, 35, 36
Aldenderfer, M.S.   140, 141

Barnett, V.   209, 210
Bartlett, M.S.   44, 51, 53, 97, 149, 153, 167
Belbin, L.   206, 207
Benzecri, P.J.   201, 207
Blashfield, R.K.   140, 141
Bumpus, H.   3, 17, 27, 28, 29, 40, 41, 46, 47, 121

Carter, E.M.   39, 40, 44, 51, 56
Causton, D.R.   26
Chatfield, C.   105, 106
Chernoff, H.   30, 31, 36
Cleveland, W.S.   36
Collett, D.   126, 127
Collins, A.J.   105, 106

Darwin, C.   1
Digby, P.G.N.   74, 75, 141, 206, 207
Dixon, W.J.   39, 56, 104, 106, 126, 127, 166, 167, 210
Dunn, G.   141, 145
Dunteman, G.H.   88, 91

Ehrlich, P.R.   17
Everitt, B.S.   32, 36, 141, 145

Fisher, R.A.   109, 127, 201, 207

Galton, F.   3
Gauch, H.G.   205, 207
Giffins, R.   166, 167

Gordon, A.D.   140, 145
Green, E.L.   157, 167
Greenacre, M.J.   206, 207

Harman, H.H.   97, 99, 106
Harris, R.J.   111, 127, 150, 166, 167
Hartigan, J.   140, 145
Hawkins, D.M.   209, 210
Higham, C.F.W.   10, 17, 53, 56, 62, 75, 89, 141, 186
Hill, M.O.   205, 206, 207
Hirschfeld, H.O.   201, 207
Hosmer, D.W.   126, 127
Hotelling, H.   3, 39, 40, 43, 48, 53, 76, 91, 146, 147, 167

Jackson, J.E.   88, 92
Jolliffe, I.T.   88, 92

Kaiser, H.F.   97, 101, 106
Kempton, R.A.   74, 75, 141, 206, 207
Khatri, C.G.   56
Kijngam, A.   17, 56, 75
Kovach, W.L.   73, 75, 135, 140, 145, 206, 207
Kruskal, J.B.   170, 171, 181, 182

Lemeshow, S.   126, 127
Levene, H.   44, 45, 46, 56, 84
Lewis, T.   209, 210
Lewyckyj, R.   174, 181, 182
Little, R.A.   210
Ludwig, J.A.   74, 75, 181, 182, 198, 200, 206, 207

Mahalanobis, P.C. 62, 63, 64, 66, 67, 73, 75, 108, 111, 115, 117, 118, 209
Manly, B.F.J. 17, 56, 74, 75, 147, 167
Mantel, N. 57, 70, 71, 72, 73, 74, 75
Massey, F.J. 39, 56
McKechnie, S.W. 8, 17
Mielke, P.W. 72, 75
Mustonen, S. 166, 167

Namboodiri, K. 26
Nash, J.C. 26

Pearson, K. 76, 92
Penrose, L.W. 62, 64, 66, 67, 73, 75
Persson, S. 141, 145

Ramsay, J.O. 181, 182
Randall-Maciver, R. 5, 17
Reynolds, J.F. 74, 75, 181, 182, 198, 200, 206, 207
Reynolds, M.L. 182
Romesburg, H.C. 74, 75, 140, 145, 176, 182
Rubin, D.B. 210

Schiffman, S.S. 181, 182
Schultz, B. 44, 56
Seber, G.A.F. 105, 106, 123, 126, 127, 210
Seery, J.B. 182
Spearman, C. 93, 94, 106
Srivastava, M.S. 44, 51, 56
Steyn, A.G.W. 36
Stumf, R.H. 36

Ter Braak, C.J.F. 206, 207
Thomson, A. 5, 17
Togerson, W.S. 170, 182
Toit S.H.C. 32, 36
Tsu, Lao *v*
Truft, E.R. 36

Van-Valen, L. 45, 46, 48, 51, 53, 56

Weber, A. 89, 92
Welsch, R.E. 30, 36
White, R.R. 17
Wish, M. 171, 181, 182

Yao, Y. 40, 56
Young, F.W. 174, 181, 182

Subject index

ALSCAL-4   174, 175, 176, 181
Analysis of variance   49, 50, 51, 52, 109, 110
Andrews plot   33–5

Bartlett's test comparing covariance matrices   44–5, 51, 53
Bartlett's test on canonical correlations   149–50, 153
BMDP computer package   36, 104, 106, 121, 124, 126, 127, 137, 140, 144, 166, 167, 210

CANOCO computer package   206, 207
Canonical correlation analysis   14, 146–67
Canonical discriminant functions   108–10, 138
Chernoff faces   30–2, 35
Chi-squared distribution   49, 52, 63, 111, 120, 121, 122, 124, 150, 157
Cluster analysis   13–14, 128–45, 208
Common factors   94
Communality   95, 101, 102
Computer programs   16, 36, 53, 73–4, 104–6, 125–6, 140, 166, 181, 206
Correlation coefficient   3
Correlation matrix   26, 80, 84, 85, 101, 153, 154
Covariance matrix   16, 24–6, 64, 79, 108, 191

Correspondence analysis   15, 201–5, 206

DECORANA computer package   206, 207
Dendrogram   128, 129, 131, 132, 135, 137, 139, 140, 163
Detrended correspondence analysis   205
Dice index   69
Discriminant function analysis   13, 107–27, 128, 208
Disparities   171
Dispersion matrix, see Covariance matrix
Distance matrix, see Multivariate distances
Draftsman's display   29–30, 35, 187, 189, 190, 195, 196, 197, 198, 199, 200

Eigenvalues and vectors, see Matrix
Euclidean distances, see Multivariate distances
Examples
  boys in a preparatory school   93
  distribution of a butterfly   7–9, 14, 57, 74–5, 146, 152–7, 166
  Egyptian skulls   6–7, 13, 51–2, 64–7, 72–3, 107, 112–14, 117, 118
  employment in European countries   10–11, 14, 84–7, 99–104, 115–17, 134–8, 164–5, 167, 183

Examples (*contd*)
  grave goods in the Bannadi
    cemetery 142–3, 144, 186–90,
    196–8, 199–201
  plant species in Steneryd Nature
    Reserve 141, 144, 184–7,
    195–6, 198–9, 204–5
  prehistoric dogs from Thailand
    9–10, 14, 30–5, 53–6, 57, 60–2,
    126–7, 138–40
  prehistoric pottery goblets for
    Thailand 88–9, 206–7
  protein consumption in Europe
    89–91, 106, 164–5, 167
  reading and arithmetic tests for
    schoolchildren 146–7
  road distances between New
    Zealand towns 172–5, 183
  soil and vegetation in Belize
    157–63
  storm survival of sparrows 1–3,
    12, 13, 27–30, 39, 40–3, 46–8,
    77–8, 81–4, 107, 121–2, 183, 208
  voting behaviour of New Jersey
    Congressmen 176–81, 183

F-test 40, 44, 51
Factor analysis 12–13, 88, 93–106,
  209
Factor loadings 94, 101
Factor rotation 96–7
Factor scores 96, 97, 103, 104
Furthest neighbour linkage 131

Graphical methods 27–36
Group average linkage 131–2

Hierarchic clustering 128–9,
  130–2, 139
Horseshoe effect 205
Hotelling's T^2 test, *see* T^2 test

Interpreting factors and indices
  102, 116, 151–2, 161–3, 190

Jaccard index 69
Jackknife classification 118

Kaiser normalization 97, 101

KYST computer program 181

Levene's test to compare variation
  44, 45, 46
Likelihood ratio test on sample
  mean vectors 49, 111
Logistic regression 118–20, 126
LOTUS 53, 56

Mahalanobis distance, *see*
  Multivariate distances
Mantel's matrix randomization
  test 57, 70–2, 73, 74
Matrix
  addition 20
  column vectors 19
  correlation 72, 73
  determinant 22
  diagonal 19
  eigenvalues and eigenvectors
    23–4, 79–80, 81, 86, 100, 101,
    110, 113, 115, 148, 153, 185, 189,
    190, 191, 193, 194, 196, 196–7,
    203, 204
  identity 19, 22
  inverse 22
  latent roots and vectors, *see*
    Eigenvalues and vectors
  multiplication 20, 21
  orthogonal 22
  quadratic form 23
  row vector 19
  scalar 21
  singular 22
  square 18
  subtraction 20
  symmetric 19
  trace 20
  transpose 19
  zero 19
Maximum likelihood 105, 119
Mean vector 16, 24–5, 108
MINITAB 126, 127
Missing values 209–10
Mixtures of distributions 140
Multidimensional scaling 14,
  168–82, 190, 198, 205, 208
Multiple regression 1, 146, 147, 151

Multiple testing 37, 43–4
Multivariate distances
  between individuals 57–60
  between populations and samples
  62–4
  calculating 74, 206
  Euclidean 58–60, 60, 61, 133–4,
  135, 192, 198, 199
  from proportions 67–8
  Mahalanobis 62–3, 64, 66–7, 73,
  108, 111, 117, 118, 209
  Manhattan 195
  Penrose 62, 64, 66–7, 73
  with multidimensional scaling
  126–7, 128, 133
Multivariate normal distribution,
  see Normal distribution
MULTISCALE computer program
  181, 182
MVSP computer program 73, 75,
  135, 140, 144, 198, 206, 207

Natural selection 1, 27
Nearest neighbour linkage 130–1,
  139
Niche overlap 68
NMDS computer program 181,
  200
Normal distribution 15–16, 38, 40,
  44–5, 49, 51, 63, 112, 118, 125,
  126, 149
Number of principal components or
  factors 82, 96, 101

Ochiai index 69
Ordination 14–15, 29, 183–207
Outliers, *see* Multivariate residuals

PATN computer package 206, 207
Partitioning cluster analysis
  129–30
Penrose distance, *see* Multivariate
  distances
Presence–absence data 68–9,
  142–3, 144, 187, 189, 196, 200,
  203
Principal components analysis 3,
  12, 14, 29, 76–92, 95, 99, 134,
  147, 184, 187, 193, 205, 208, 209

Principal components factor
  analysis 97–9, 105
Principal coordinates analysis 15,
  190–5, 205, 206
Prior probabilities in discriminant
  function analysis 117
Profile plots 32–5

Quadratic discriminant functions
  126

Reciprocal averaging 202
Residuals and outliers 63, 111, 209
RT computer program 4, 75

SAS computer package 105, 106,
  126, 127, 166, 167, 182
Separate sampling 120, 122
Similarity measure 68, 192, 194
Simple matching index 69
Size and shape 88–9
SOLO computer package 36, 106,
  121, 122, 124, 126, 127, 137, 140,
  144
Spatial correlation 74, 75, 166
Specificity 95
SPSS computer program 105, 106,
  166, 167, 182
Star plots 30–2, 35
Stepwise discriminant function
  analysis 117–18
stress 171, 200
SURVO computer package 166,
  167

t test 38
T^2 test 39–40, 43, 45, 48, 53,
  110–11
Two factor theory of mental tests
  94
Type one error 43

Value of factor analysis 105
Van-Valen's test to compare
  variation 45–6, 48, 51, 53
Varimax rotation 96–7, 101, 105

WordPerfect 36